INNER EXILE

Elisabeth Heisenberg

INNER EXILE

**Recollections of a life
with Werner Heisenberg**

Translated by S. Cappellari and C. Morris
With an introduction by Victor Weisskopf

BIRKHÄUSER
Boston • Basel • Stuttgart

Originally published under the title "Das politische Leben eines
Unpolitischen — Erinnerungen an Werner Heisenberg."
© R. Piper & Co. Verlag, Munich 1980

Birkhäuser Boston would like to thank
Irene Heisenberg for her assistance and advice.

Library of Congress Cataloging in Publication Data

Heisenberg, Elisabeth, 1914-
    Inner Exile.
    Translation of: Das politische Leben eines Unpolitischen.
    Includes index.
    1. Heisenberg, Werner, 1901-1976. 2. Physicists —
Germany — Biography.   1. Title.
QC16.H35H4413   1984   539.7'092'4 [B]   83-10023
ISBN 0-8176-3146-1 (Switzerland)

CIP-Kurztitelaufnahme der Deutschen Bibliothek

Heisenberg, Elisabeth:
Inner exile: recollections of a life with Werner Heisenberg /
Elisabeth Heisenberg. Transl. S. Cappellari and C. Morris.
With an introd. by Victor Weisskopf. - Boston; Basel; Stuttgart:
Birkhäuser, 1984
    Einheitssacht.: Das politische Leben eines Unpolitischen [engl.]
    ISBN 0-8176-3146-1

© Birkhäuser Boston, 1984
ISBN 0-8176-3146-1
Printed in USA

A B C D E F G H I J

He was first of all a spontaneous person, following that a scientific genius, next an artist close to
the creative spark, and only in the last instance,
out of a feeling of duty, a "homo politicus."

Carl Friedrich von Weizsäcker
on Werner Heisenberg

CONTENTS

INTRODUCTION

In the years between 1924 and 1927 some of the deepest riddles that nature posed to us were solved: how to understand and describe the structure of atoms and, therefore, the structure and behavior of matter, since all matter is made of atoms. It was a truly revolutionary step, because it required the abandonment of many old concepts and prejudices and the creation of new concepts and a new language called quantum mechanics, in order to understand and describe what happens within and between the atoms. A new subtle reality was discovered to exist in this realm, on which the ordinary reality of our daily life is based. The new insights were achieved not by any single individual, but by a small group from different nations, with Niels Bohr in Copenhagen as the most powerful leader. Most of these people were very young, in their twenties, whereas Bohr was in his forties at that time. It was a little group of enthusiastic young spirits, well aware of being at the front line of knowledge, of shedding light on a previously murky and contradictory situation. Never before have so few contributed so much insight into the workings of nature in such a short time.

One of the young men in this group was Werner Heisenberg. He was perhaps the most active and creative among them, the one who provided the most important ideas and formulations. He and the others did their work mostly at their home universities, but they spent much time in Copenhagen or had a vivid correspondence with Niels Bohr and among themselves. For many of them, especially for Heisenberg, Niels Bohr was a father figure; he provided encouragement, criticism and the philosophical view that was necessary to understand, interpret and accept those new ideas that, even today, are difficult to grasp.

Heisenberg had a special intuitive way of getting to the essential point. This, together with an incredible force of persistence and determination, made him the most prolific and successful physicist of the recent past. Whenever important problems turned up in the subsequent development of quantum mechanics, more often than not, it was Heisenberg who found the solution. He pointed to the direction of further developments by inventing new ways of looking at the situation. Apart from his fundamental contributions to the formulation of the quantum mechanics of the atom, he was able to decipher the helium spectrum that had puzzled the physicists for decades; he explained the magnetism of iron and similar metals; he paved the way to get a profound description of nuclear structure by considering the proton and the neutron as two states of the same basic particle. These are only a few of his outstanding contributions. All of them contained seminal ideas which led research into new directions and found their way into the foundations of physics.

The most quoted of his contributions is the Heisenberg uncertainty relation, which is the basis of the new understanding of atomic reality. It defines the limits to which the ordinary classical concepts such as "particle" and "wave" are applicable. Beyond these limits the new subtler reality of the quantum state emerges. Heisenberg's discovery and formulation of these fundamental limits is a typical product of the collaboration of Niels Bohr with his younger colleagues. It was largely the result of numerous discussions, extended over long periods, in the institute, in letters, on excursions along the Danish coast, or in the German Alps. Bohr, in his Socratic way, asked the relevant questions and pointed to the depths of the problems and the plentitude of the phenomena. Then in the minds of the best of his disciples, the new ideas grew and took shape. Obviously, such a relationship must bring along a deep personal friendship, a sharing of philosophy and common outlook, and many other human, artistic, and literary concerns. So it was between Bohr and Heisenberg until . . .

Heisenberg was a pleasant man who made friends easily. Everybody who came in touch with him liked and admired him. He was interested in all sides of human culture; he was an accomplished pianist and an enthusiastic mountaineer and skier. So one would have thought that he would have had a happy and fulfilled life as a leading scientist, in the midst of many students, a life devoted exclusively to a deeper understanding of physics and the enjoyment of art, music and literature and the beauties of nature.

It did not come that way. He and his contemporaries lived in times of political upheaval. Europe, and in particular his homeland, went through one of the worst periods in world history. The Nazi regime unleashed the most destructive side of human nature, antithetic and opposed to the cultural values cherished by Heisenberg and the group he belonged to.

The upheaval in Europe caused untold horror and destruction. Millions of people lost their lives, millions had to live under the cruelest oppression and humiliation; the minds of western civilization were perverted to such a degree that it still has its aftereffects 40 years later, in having weakened our sensitivity toward human life and oppression. Those who survived the holocaust should consider themselves lucky; most of them do not think about it anymore. [Our present time has its own madness, the nuclear arms race and the threatening nuclear war.] They suffered in various degrees from the cruel acts of Nazi oppression, from the mild form of forced emigration to the cruelest form of corporal and mental torture and humiliation in the concentration camps. Heisenberg and his family were not exposed to any of these punishments. They did not suffer in this direct sense of the word. Compared to the fate of those victims of Nazism, their burden was light.

Elisabeth Heisenberg's book is important because it provides insight into some aspects of life under an oppressive regime that are not often recounted. Here was a man deeply rooted in the best of German cultural life. It was in the German intellectual tradition to devote oneself to the "higher" concerns and to leave the dirty game of politics to lesser spirits. One assumed that those political uproars would somehow go away and the better part of German conscience would win out. Heisenberg was looking for positive signs in the national movement, but he saw its true nature and was deeply repelled by its spirit, by its actions and excesses.

What then were his choices? He could emigrate, he could actively participate in the underground movement, or he could retire from public life and live as decently as possible with his family and his work. The last possibility existed for many other Germans but not for him. He was much too influential and prominent. The participation in the underground movement required an amount of heroism that one cannot ask from any person. Heisenberg was the head of a family of six children. He was not the heroic type, but rather, a careful and prudent character. An active role in the underground could not have been his way.

It is one of the aims of this book to explain why he did not choose to emigrate. Many friends abroad strongly encouraged him to do so and offered him, of course, the best possible opportunities. His life and that of his family would have been much easier. But he thought it would be a cop-out. He was deeply attached to Germany and he felt a responsibility toward those who shared his feelings and whom he might be able to help because of his influence and reputation.

Was there another way? The answer to this question is the main concern of Elisabeth Heisenberg's book. It was a hard way, not in the sense of bodily suffering — he and his family were spared that — but in the daily struggles of conscience and the fear of being found out as a doubter, or even a traitor. It must have driven him to utter despair and depression that his beloved country had fallen so deeply into the abyss of crime, blood and murder. He learned the truth of what Max Planck had told him: "In the ghastly situation in which Germany now finds itself, no one can act entirely decently." He "envied those of (his) friends whose lives in Germany had been made so impossible that they simply had to leave." He wanted to stay in Germany and to help create "islands of decency" where, within small groups, some of the cultural achievements might be saved to serve as the beginning of a new cultural life when the Nazi holocaust was over.

The question was raised whether Heisenberg believed in a German victory before the turn of events. There were people who had heard him say so. But his book says definitely that he never did. This is not necessarily a contradiction. Heisenberg was very prudent and circumspect when he talked to others during the Nazi time. Those who never lived under a dictatorship may not comprehend the tremendous danger of being given away, the fear for their life and the lives of their family. A report back to the Nazis that he doubted victory could have been a death sentence. Even if he were a hero—he was not—he would not risk his life only to improve his reputation among his acquaintances.

Obviously, by remaining in Germany, he was soon involved in the efforts of the Nazi regime to exploit nuclear energy. The possibility of a bomb was evident. There was a vast difference between the position toward the bomb problem held by the scientists on the allied side and by those on the German side who were more or less opposed to the regime. Heisenberg was certainly one of the latter. The allied scientists were terrified by the idea of a bomb in the hands of Hitler; they trusted the policies of their leaders, Churchill and Roosevelt. Those German scientists would have been in a terrible situation in devising this awesome weapon, torn between distrust for Hitler and the pressure to collaborate with the Nazis. Fortunately for them and for the world, something happened that helped them out of this dilemma. The estimate Heisenberg and his collaborators made of the time and effort necessary to devise and construct the bomb was set at several years, and the effort was considered to be beyond the German industrial capacity during the war. So the Nazi government decided to give it up and to concentrate only on the power production as a long-range project, of questionable use during the war, but of promise for the future.

It is improbable that Heisenberg and his colleagues consciously exaggerated the time and the effort in order to avoid the possibility of giving the weapon to Hitler. They probably expressed their true opinion. But in such situations nobody knows how much he is influenced by his willingness to do it. Every great project can succeed only if its proponents support it with full conviction of its necessity. Indeed, such conviction usually leads to an underestimation of the time and effort; without such underestimations, few daring projects would ever have been begun and accomplished.

The negative bomb decision saved Heisenberg and his colleagues from many troubles; Nazi authorities would have breathed down their necks day and night, unavoidable setbacks would have been interpreted as treason and the responsible people persecuted. Furthermore, Heisenberg and his group would have become schizophrenic between the pangs of their conscience and their sense of duty to the nation. So a stroke of luck enabled them to retire to a reasonably well supported government side effort, a haven for saving people from service at the crumbling front lines. They considered their work, not without justification, as a basis for the start of a nuclear industry in the postwar world.

One tragic point needs to be discussed. It is the unfortunate and abortive visit of Heisenberg with Niels Bohr during the war. A great friendship and a creative human bond was shattered. Certainly, an end to a friendship cannot be compared to the mass murders and the effects of mass bombings, but it is a symbol for the tragedies of war. Heisenberg went to Copenhagen, to his old friend, mentor and father figure in order to talk to him about the great problems that the nuclear explosive had created for mankind and in particular for the community of scientists. But he did not consider how deep and justified the feelings of hate and desperation were in a nation victimized by the Nazis. He expressed himself very vaguely, fearing that any direct statement about Germany's nuclear effort, or any doubt of a German victory, would put him and his family in mortal danger. Under these conditions, it was difficult for Niels Bohr to see in Heisenberg only the old disciple and friend and not a representative of the oppressors. Perhaps one could have expected that Heisenberg would have spoken openly and clearly with Bohr at a time when they could not have been overheard. Here it was not a question of what others might think of him. He faced his fatherly friend who would have taken all precautions and would never have given him away in spite of his misgivings, knowing the consequences only too well. Perhaps such an open exchange would have been useful.

Whatever has happened, nobody has the right to reproach anybody for having avoided a deadly risk. Who can stand up and say he would have taken the risk in the same situation with the same responsibilities? Those who have never been in such situations must be grateful to their fate of being spared such decisions.

In times of oppression and persecution, the general principles are less important than individual personal actions. We know that Heisenberg did what he could to protect the Bohr Institute after the German occupation. He saved the lives of a number of people. He never made much of it, he did not mention it in his writings, but we hear about several such cases in this book, and there was a letter published in the magazine, 'Science News,' [Vol. 109, p. 179, 1976] by one of those whose life was saved by Heisenberg. Such acts weigh more than any statements.

Heisenberg believed in the importance and strength of the international community of scientists. He had become acquainted with it during the early years of quantum mechanics research in Copenhagen. He saw the great edifice of thought emerge from a group of human beings of very different national origin, but united by their common enthusiasm to lift a part of the veil behind which the secrets of nature were hidden. He was disappointed that this spirit of community did not withstand the divisive forces of the World War.

When he was free again, after the war, to devote part of his efforts to fundamental problems of physics, his intuition enabled him to contribute again some important seminal ideas about the structure of elementary particles. But in the postwar period two unsuccessful attempts cast a shadow over his life and robbed him of some of his optimism. One was his failure to establish an effective board of scientists advising the government on essential questions; the second was his failure to convince the world of physicists of the power of his last attempt to formulate an all embracing theory of matter. He may have overextended himself in both attempts. His greatest source of success was always his ability to forget about details and aim directly at the essential points. Perhaps at the end of his life, the plentitude of events in physics as well as in public life had become too overwhelming. This may be one of the predicaments of today's world in general.

It is hoped that his doubts about the future of the scientific community were exaggerated. After all, the international spirit of science is not at all dead today. One of the most significant examples is CERN, the international laboratory in Geneva which he helped to bring about. Not only is it run by twelve European nations, but under its roof scientists of the whole world collaborate, irrespective of political differences. We find Americans, Russians, and Chinese working together on the same experiment. There exists a strong urge among physicists to establish better bonds and better understanding between the politically divided parts of humankind. Perhaps this urge comes to some extent from a feeling of guilt: that some of the great insights of their research have been exploited to serve as weapons of annihilation and as a threat to the future of mankind.

Elisabeth Heisenberg's book serves as an important human document. It describes the difficulties, tribulations, and conflicts which people encounter under a regime of dictatorship and oppression even when they are not direct victims of the system. It describes the hardship and mental sufferings of a mother and wife who accepted the decision of her husband to choose a thankless and difficult path, and helped and supported him as much as she could. Great were the sorrows and fears, but great also were the moments of joy and gratitude when things worked out better than they had feared. The book shows us how complex and how exasperating life can be when you are forced to compromise with, and adjust to, intolerable conditions in order to save what you consider to be more important. Such regimes still exist all over the world. That is why this book will also enhance our understanding of the predicaments of scientists in similar situations today. It will help to improve the bond between those who suffer and those who have been spared. In this sense, the book will contribute to the strengthening of the international community of scientists.

V. Weisskopf

April 1983

PREFACE

In the spring of 1977 I travelled through Germany with about 25 recipients of the Humboldt Scholarship. Trips like these are granted to young scientists by the Alexander von Humboldt Foundation. The students come to Germany from over 50 different countries to work and continue their studies; during their visit they are offered the opportunity to become acquainted with Germany. For this particular trip I had proffered myself as a guide for the young scholars. It turned out to be an unforgettable experience, with most stimulating encounters. During the course of this three-week journey we naturally had occasion for many conversations about the past and the calamitous history of Germany. After all, I do belong to the generation that had to endure and, in a certain sense, also had to share the responsibility for this time. Naturally, these discussions turned to the question of the role Werner Heisenberg, my husband, had played in these events, and, furthermore, why he had remained in the country. To me, this revealed a keen interest. However, the various, sometimes almost abstruse ideas of my husband's role that surfaced in the course of these discussions made me aware of how hazy, distorted, and often quite false and contradictory people's notions were. And so, the idea began to take shape to write down, at some point, what I know about Werner Heisenberg, and to draw the real picture as it has emerged over the many years during which I shared his life in a constant exchange of thoughts, and as it lives on in me now. Thus I decided to describe Heisenberg's political life and the struggles the criminal history of his country had subjected him to. There were to be no embellishments; nor did I want to hide his weaknesses or his mistakes. But I also wanted my account to convey the unfailing consistency and integrity inherent in Heisenberg's life. These qualities are known to all who were close to him.

I was encouraged to pursue this work by a friend and colleague of Heisenberg, who had been forced to leave Germany by the National Socialist myrmidons and had emigrated to America. He said to me: "Yes, write it! Only you can revitalize the image of the person Heisenberg, so that it will convincingly match with the image the world already has of him as a pioneering scientist." This gave me the courage truly to follow through on these notes. Essentially, they are memories of Werner Heisenberg, of what he told me, of our common experiences, of conversations that, due to their unusually daring content, have stayed in my mind very clearly. Sometimes I also refer to what others have reported, sometimes to letters he wrote at a later date. But it would be a mistake to assume that I have conclusive new documents capable of shedding a completely new and clear light on the image of Heisenberg. This is not the case. And with this, I already touch on a fundamental problem, the significance of which is often not correctly perceived, and on occasion is actually underestimated — a problem that leads us deeply into the troubles of those times. For it is within the essence of a regime of terror to convince the individual that he dare not articulate his political thoughts and feelings [especially if they contradict the official view], let alone write them down in letters or diaries. To do so would be to put one's life, or at the very least one's freedom, in jeopardy. This makes grasping the political truth, the motives of individuals living and interacting under the Nazi rule, so difficult; indeed, this approach all too easily leads to simple truisms and uncritical generalizations.

For all those who did not live through those times and

who know about them only in retrospect, be it from books or the reports of others, these words might possibly sound somewhat dubious, as though I were trying to hide or gloss something over. The course of events is known: the dreadful reality of what happened is always present when thinking of this period. Nonetheless, it is difficult to imagine how it really was, and what the day-to-day life of the individual was like under such a thoroughly evil system, perfectly organizing terror and seduction, with its targeted lies and diabolically measured horrors aimed at intimidating the population. From our current vantage point it is impossible to imagine the lives of those not sympathetic to this regime and not wanting to have anything to do with it, but who, in spite of everything, did not fully capitulate. Life went on, even though the terror and anxiety were real. We all had to disguise our words and actions as best we could. Even among acquaintances and casual friends we were guarded, and we spoke only in innuendoes. Who was to know which casual words might be repeated where? And if something was said that was not immediately clear, an uncanny, oppressive hush took hold of the conversation, and an uneasy feeling lingered on. This, precisely, was the evil intent of the Nazis — to sow anxiety and distrust and thereby to poison human relationships, frequently even destroying their very foundation. This silence burdened with rejection and mistrust, this sullen, defensive way of relating became a trademark of the time.

But how much more dangerous it was to commit one's political thoughts, feelings, and opinions to paper! Everything that today could prove Heisenberg's unequivocal political opposition to the Nazi system could have cost him his life at the time had it been documented in writing. Heisenberg was completely aware of this. He was careful; after all, he had not stayed in Germany to die as a martyr. He had no intention of perishing in the senseless killing — he intended to fight for his life. His reasons and motivations will be discussed further on. The above, however, explains why there is no new, irrefutable evidence to substantiate Heisenberg's political alibi. His life should be evidence enough.

Naturally, there are letters written by Heisenberg during this period that have a political content and that reveal his often depressed and desperate state of mind. Some of these letters were sent to Sommerfeld, his teacher in Munich, and some to Niels Bohr. However, they can always be interpreted in a purely personal manner. At the time, there was even an exchange of letters with Sommerfeld discussing the catastrophic development of the policy regulating the sciences. In this matter, he was completely frank; nevertheless, even here he managed to avoid taking any generally binding position. He was convinced that his letters were being opened and read. Thus, he often deliberated the exact phrasing of a sentence for a long time, to be sure it communicated precisely what he wanted it to, without its being able to compromise him. I still clearly remember an instance of his reading a letter to me; its formulation had given him grim satisfaction, knowing, as he did, that his surveillants would be unable to decipher a thing, whereas the recipient would know exactly what had been said. But even thus forewarned it is still difficult to discern Heisenberg's real political opinion from his letters—they were devised to avoid precisely that.

And so, I do not have much more to offer in these notes than my memories, and the inner consistency and harmony of this man's life. They transmitted themselves to me ever more forcibly and clearly as I wrote this down, and I hope they will have the power to bear their own witness.

Several of Heisenberg's friends were opposed to my work. They were of the opinion that everything essential had already been said in Heisenberg's autobiography *Physics and Beyond*, including the clarification of his political attitudes. Why, then, another book? His book should be read in order to discover how Heisenberg guided himself through the labyrinth of our complicated world. What I want to say here is entirely different from what he wrote about himself. Heisenberg was bashful. He was utterly incapable of standing up in his own defense, and he also believed it to be unnecessary. He meticulously guarded his personal life from public scrutiny. Thus he discusses his personal affairs in vague terms and with large gaps in his autobiography; it had never been intended to be an autobiography in any case. Indeed, in his book Heisenberg was not writing about himelf; he was writing about his thoughts on science, politics, philosophy, and religion. And in the book he traces the development of these thoughts and the sources from which he had derived them. The texts are refined extracts from a fully experienced life as a thinker. But very little can be learned about Heisenberg himself, about his personal conflicts and problems. He thought only one thing to be important: he wanted to communicate what he had recognized to be the truth — and that is what he did in his books. But he did not think his personal life significant enough to make much of a stir about it. For this reason, his book leaves certain aspects open, creates opportunities for many a false speculation and distorted opinion, but also for curiosity and a justified interest in his personality, the image of which fluctuates so oddly between awe and criticism. As his friend, the Dutch Professor Kramer, once replied, when Heisenberg, in a somewhat disturbed and uncertain frame of mind, queried him in this matter: "Well, Heisenberg, you know your Schiller, after all; as it says in the prologue to 'Wallenstein's Camp,'

Twisted by the hatred and goodwill of the factions,
The image of his character wavers in history...

So you, too, will be remembered by future generations!" he added, and laughing happily, they parted again.

I add my notes to the open questions lingering in

Heisenberg's book, and talk about his life, the person
Werner Heisenberg, who was loved by everybody close to
him, and who himself was positive about his life, and thus,
in the end, could take leave of it with great peace of mind.
These notes, as well, can only make a small contribution
to explaining the man; much still remains unanswered. This
will probably never change, for it is difficult to fathom the
many levels of this multi-faceted person. I speak here only
of Heisenberg's conduct in the political world, the world
that intruded into his life with such great animosity. This
controversy with politics quite necessarily determined a wide
area of his life and, together with his scientific work, took
up a great part of his ideational world. To be able to understand his political actions, it is necessary to see the widely
dispersed surroundings in which his personality was rooted,
and out of which his political views were conceived, grew,
and matured. For this reason I will also talk about his childhood, the home of his parents, his friends, music, and his
beloved country, which he thought more beautiful than anything he witnessed on his travels through the world; for
he felt he belonged there, his roots were in his landscape.

SOURCES

Naturally, I have not been able to recreate the entire content of this book from my memory and from letters. To Dr. Ballreich, I offer thanks for much material on the rebuilding of the Kaiser Wilhelm, i.e., Max Planck, Society. He is in the process of writing a history of the Society and has collected a rich variety of material; in several letters he placed many an interesting detail of his knowledge at my disposal.

Furthermore, I drew upon several facts from the description of Armin Hermann,[1] and utilized one or two passages from a manuscript of his, which later appeared in the book, *Die Jahrhundertwissenschaft*,[2] in an abbreviated version. The quote of Albert Speer was taken from his memoirs.[3] I obtained very valuable information from Herbig's book, *Kettenreaktion*.[4] In America I read Goudsmith's book, *Alsos*,[5] and the pertinent passages in the book by General Groves.[6] In addition, Beyerchen's book, *Scientists under Hitler*,[7] came to my attention, though I do not quote from it.

Direct quotes of Heisenberg are derived either from his lectures[8] or from his book, *Physics and Beyond*[9] — some, however, I took from letters. The origin of the Weizsäcker quotes is partially in unpublished speeches.[10]

I offer my grateful thanks to all who have helped me during the three years it took me to write this book, who encouraged me not to let it slip into oblivion in some forgotten drawer. I would especially like to name Nellie and Kurt Friedrichs of the Courant Institute in New York, with whom I discussed the manuscript for countless hours; Professor

Sources

David Nachmansohn of Columbia University in New York, who supported me in the decision to let this work ripen into a book; Professor Victor Weisskopf of Cambridge, Mass., who encouraged my efforts to write; and, likewise, Professor Ben Kedar in Jerusalem.

Finally, I would also like to thank my supporters, Professor von Weizsäcker and Professor Lüst, and, last but not least, Dr. Bohnet, with whom I went over the manuscript once again to our mutual enjoyment. Without all this help, this book would never have taken shape. And now I send it out into the world, like a ship being launched to face the winds of its destiny.

E.H.

Munich, May 1980

REFERENCES

1. Armin Hermann, *Heisenberg*, Reinbek, 1976.
2. Armin Hermann, *Die Jahrhundertwissenschaft*, Stuttgart, 1977.
3. Albert Speer, *Inside the Third Reich: Memoirs of Albert Speer*, New York: Macmillan, 1970.
4. Jost Herbig, *Kettenreaktion*, Munich, 1976.
5. Samuel A. Goudsmit, *Alsos*, New York: Tomash Publishers, 1947.
6. Leslie R. Groves, *Now It Can Be Told: The Story of the Manhattan Project*, New York: Da Capo Press, 1962.
7. Alan D. Beyerchen, *Scientists under Hitler: Politics and the Physics Community in the Third Reich*, New Haven: Yale University Press, 1981.
8. Werner Heisenberg, *Schritte über Grenzen*, Munich, 1971 (new, expanded edition, 1973).
9. Werner Heisenberg, *Physics and Beyond*, New York: Harper & Row, 1971.
10. Carl Friedrich von Weizsäcker, *Werner Heisenberg*, Munich, 1977.

chapter one

CHILDHOOD AND YOUTH

Heisenberg's youth played an important role in the development of his political conduct. He himself always claimed that the events of childhood and youth, and the ideas fostered in the course of them, are the fundamental and most strongly influential forces of a lifetime. They would be called upon again and again. Without doubt, in his case this was true. The diverse structure of his family, the war, the revolution, and the youth movement were the crucial experiences that shaped his social and political understanding.

Werner Heisenberg was born in Würzburg on December 5, 1901. Thus his entire childhood took place under the impression of the period prior to the First World War, the time of imperial Germany and the Kingdom of Bavaria, in which he grew up. Among the few valuables he possessed with a certin demure pride was a pair of gold cuff-links decorated with a crown and a lavish 'L.' When he was a small boy of about eleven, they had been presented to him as a gift by the Prince Regent Ludwig [later King Ludwig III], when, on the occasion of the latter's visit to the Max-Gymnasium in Munich, he recited a short poem composed by his mother in honor of the royal guest. His mother was the daughter of the school's headmaster and it was owing to this, as well as to the charm of her poetry, that little Heisenberg had been accorded the honor, and his clear, sharp eyes had, no doubt inspired the favor of the sovereign. Heisenberg's mother blended intelligence with a loving heart that had never really matured. Under the sway of an autocratic father and the stormy temperament of her husband, she had, like so many other women of her generation, never achieved mental independence. But she had paved the way for a happy childhood for both of her sons; Heisenberg felt secure in the circle of a stable family. He remembered his mother mainly as the protective figure at whose side he took his first steps and caught his first glances of a fascinating, colorful world, and he was always indebted to her in love and gratitude. One of the earliest memories he often recounted was of the colorful panes of stained glass in the *Käppele*, a church of pilgrimage on a hill overlooking Würzburg;* through these he could observe the world stretched out below him in a new light and in changing colors. He never tired of witnessing this phenomenon, and it still occupied his awakening spirit for a long time to come.

His one-and-a-half-year older brother was the trusty playmate of his childhood, although they did have their occasional altercations to overcome. In spite of his predominantly peaceful nature, Heisenberg could become quite angry, even fly into a rage — he later called this his 'bloodlust,' though it was actually Otto Hahn who first coined the phrase — and as a result, the fights between the brothers could sometimes assume quite violent forms. One day, following a battle with chairs that left them both with painful injuries, they decided to forsake this kind of fracas once and for all; this decision was characteristic for both of them. At the time they were probably 13 and 14 years old, and in the end, they both determined that it was nonsensical and stupid to try to settle their disputes in such a manner. They had taught each other that conflicts can be solved more successfully and reasonably through peaceful means.

The boy grew up surrounded by a world that was still intact. He felt loved and protected in the far-reaching circle of his family. This, coupled with a happy disposition, led to the development of a deep trust in people: he was always willing to see their good side first. Throughout his life he never abandoned this attitude, despite all the iniquitous and

---

*I was unable to find the colorful glass panes when I traced this childhood experience.

terrible experiences he had to face. This constant good faith helps to explain the strong reaction, almost over-reaction, to accusations or, even worse, to punishments he thought to be undeserved, that was a characteristic of his. He would retract into himself, into his own world, where he was content, and break off every contact to the person he felt had treated him unjustly. He was never confronted with this sort of problem by his parents; their relationship always struck me as being singularly untroubled. But, as it will, it did occasionally happen with more remote acquaintances. When, for example, during his first year of school, a teacher once struck his hands with a rod so hard that his fingers swelled up, hurting him badly — and this for some petty reason he did not understand at all — he never looked at the teacher again, and refused active cooperation with him. This episode is noted in the conference briefs of the school in Würzburg, where it was recorded with great concern. On the other hand, he opened up freely and without guile if his trust was not abused. He would become cheerful, was full of ideas and whimsical jokes, and was willing to help without a second thought.

No doubt it was his own great sensitivity that taught him, during the course of his life, to deal with his fellow human beings cautiously. In general, he responded to them with openness, good will, and trust; if his expectations were disappointed, however, he reacted with irrevocable repudiation, breaking off all ties once and for all. This only happened rarely and was always the exception rather than the rule. Indeed, the care and the kindness with which he treated people increasingly transformed him into a fair, patient listener and counselor; he inspired trust in both simple and privileged people alike, in workers in their shops, as well as in industrial managers and politicians. He had the gift of gaining people's confidence and of instilling trust.

When Werner Heisenberg was eight years old, the family moved to Munich. From that point on, the city was his declared and beloved home. The move brought an end to the first phase of his life; a new, more independent and expansive phase was ushered in. True, the family now lived in a flat in one of the large apartment buildings on the Hohenzollernstrasse in Schwabing, and the young boy sorely felt the "confinement" as a loss; he missed the streaming river with its life and its mysterious width, the soft green hills and the nearby forest — the Luitpold Park was only a meager substitute. But another, more colorful life had now begun as his horizons expanded in many directions: new people, new stimuli for his inquiring mind, new interests and projects. Since learning came easily to him, he had enough time to pursue these projects. He liked to build things; together with his brother, he built a large battleship about three feet long, equipped with tiny, handmade guns that could actually be fired and were electrically triggered. For the time it was a small technological marvel. He showed it to me with childish pride when I first visited his parents' home. But more than that, it was now music that imbued his life with an as yet unknown intensity. He had quickly achieved a certain level of accomplishment on the piano, permitting him access to the great works in the literature of music at the early age of 13 or 14. And since he sight-read from a sheet with great assurance, he was soon sought after for his abilities as a chamber musician; duos, piano trios and quartets unlocked their nearly inexhaustible riches to him. During these years he frequently contemplated becoming a musician.

Basically, however, the path of his future life had already been charted for him, for he moved about in the world of numbers and their internal laws, in the wide realm of scientific ideas — which he could assimilate with almost playful ease — with unflagging fascination. "I was something of a child prodigy," he told me, just a little sheepishly. At the age of 14 he prepared a friend of the family, who wanted to take her doctorate in chemistry, for the necessary mathematical exam.

But before he was 13, the First World War broke out, putting a quick and palpable end to his happy, carefree childhood. As a captain of the reserve, Heisenberg's father was immediately drafted at the outbreak of hostilities; he went off to war — at least in the memory of his son — just as it can sometimes still be seen in old picture books: on a mighty horse of war, wearing a gleaming helmet, a handsome uniform, and a long sabre at his side. His father had received orders to protect Osnabrück, his home town where he had served in the army, from enemy aircraft. A soldier with an assembled machine-gun was posted on the roof of the tallest building in the city with the job of shooting down any approaching enemy planes. This was the look of things at the beginning of the First World War.

The family accompanied the father to Osnabrück. On this trip the 12-year-old boy was greatly agitated by his experience of the enthusiasm bursting forth from people marching off to war everywhere. The scene was the same at every stop: exulting masses of people, crying women, singing men covered with flowers. Later, Heisenberg often described to me how unearthly, enigmatic, and exciting this all seemed to him. Weren't these people marching off to war, into death and destruction? He felt the contradiction. But he also clearly perceived the enhanced feeling of being alive that had gripped these people, their ready devotion to something larger than their own small, narrow lives, something that lifted them out of their daily drudgery and set them down in a world of larger relations, relations they themselves did not comprehend, but that became vaguely outlined by the rallying cry: "For Kaiser and Country!" Heisenberg repeatedly pondered this experience, philosophized about it. He even writes about it in his autobiography, *Physics and Beyond*. In a certain sense, it became one of the key experiences of his whole later life.

Only too soon the deadly and destructive face of the war became a self-experienced reality for young Heisenberg, for his beloved cousin from Osnabrück, his only slightly older best friend and playmate, had volunteered for front-line duty with this very enthusiasm, and a short time later the family received news of his death. And the elder brother of this cousin [he, as well, was a close childhood friend], who had been drafted and cheered on with flowers when he set off, returned home as a completely different person on his first leave. His bright spirits had been darkened by horrible sights. Young Heisenberg was deeply shocked by these incidents; skepticism was irrevocably implanted in his mind, and a development was initiated that led to the question whether there could possibly be a political goal capable of justifying such sacrifices. Heisenberg was never a real pacifist, but he abhorred war with all his heart. He did not believe that war could be stamped out, that a world without war could exist. He was convinced that a pacifist was doomed to failure from the start, later mentioning Einstein to me in this context, who had been a truly committed and ardent pacifist, but who, in spite of this, had finally encouraged Roosevelt to build the atomic bomb as a weapon to be used against the Germans.

Heisenberg rejected heroic tales about soldiers as false, but he also refuted the cynicism often used to represent soldiers as no more than pitiful clods, stupid enough to be abused by those in power. This image infuriated him, for his personal experience had been more to the contrary, and he understood the tragedy of war quite early: the tragic antithesis between the person genuinely putting his life at stake and, in a certain sense, transcending himself through his willingness to sacrifice and suffer, and the appraising military and political calculation that operates with, and all too often coldly exploits, this willingness. This antithesis climaxed in a horrid perversion in the Second World War.

The conflict this problem confronted him with sharpened his capacity for judgment. The powerful emotions called forth by the onset of war and its oppressive, terrifying reality continued to affect him, and taught him to differentiate between the individual's intensified sense of being alive and uncritical and dangerous euphoric states; these he detested deeply, making him completely immune to all the mass hysteria and mass seduction he later witnessed, when Hitler lured the masses into his takeover. But he had also comprehended that only a complete personal effort can bring about a deepened intensity of living; and this he thought to be better and more worth striving for than the most enticing sensual pleasure. The following lines from Schiller's 'Wallenstein's Camp' were always one of hs favorite quotes: "And if you do not put your life at stake, your life will never be yours to make." During the course of his life, this feeling was embodied for him in the dictum "What you do, do with all your strength; only then can something worthwhile come of it and truly make you happy." How often I heard him say that! In his scientific work he expressed it this way: "You just have to be able to drill in very hard wood, as well, and keep on thinking beyond the point where thinking begins to hurt." Time and again, this made the quickening of life's intensity accessible to him, even when he had become old.

In general, Heisenberg experienced the war as a time of severe deprivation, a time of hunger and need. At the age of sixteen, debilitated by exhaustion and weakness, he crashed into a roadside ditch with his bicycle. He decided to sign up for civilian war service as a laborer on a farm in Miesbach, a small town in Upper Bavaria. He had to work hard for an extremely simple fare, but he managed to recover his strength. The new experience he gained there as a farmhand left him with a strong impression. "I learned how to work there!", he would often impress on his children in later years. He learned how to exert himself, how to endure, how to use his own energies most effectively and rationally, so as to achieve the highest possible efficiency. From then on he was no stranger to hard, physical labor, he even liked it; and later on, living in his cottage on the Walchensee, it caused him pleasure and a feeling of physical well-being to fell trees and work them into firewood with his own hands. From that time on he was above intellectual pretentiousness, and the prejudices of the bourgeoisie toward the working class no longer existed for him.

The time spent on the Grossthaler farm in Miesbach would certainly have been less fruitful, and would probably have remained an insignificant episode, had his own family not had its roots in a social class quite different from the middle class-academic world Heisenberg grew up in in Munich. The Heisenbergs came from Osnabrück, from a family with a long tradition of artisans. The grandfather was a sound, respected master locksmith, who lived in a small house on Lohstrasse. His wife came from a farm in the surrounding area. As a child, Heisenberg often spent his vacations in Osnabrück. He loved this unadorned world; it was his second home. He admired his grandfather, a peaceable and relaxed man who always managed to have fine, white hands when his family gathered, in spite of his coarse work: this struck Heisenberg as a mirror image of the man's true nature. He also loved the family centered around his grandfather's home, the solidity and the steadiness of their simple world, the unbending values governing their life, a life without excess, of thrift and industry, but appreciated with warm comfort and filled with simple joys. The sister of his father, who managed the gardens of the Osnabrück castle, exercised a great influence on him with her serene and firm dignity. He was deeply anchored in this world, and as long as it continued to exist, he felt profoundly attached to it. These ties doubtlessly helped to impress upon him the ideas and convictions that were later to shape his political decisions.

To fully understand the political soil that nourished him,

however, it is necessary to say more, especially about the father, who, in spite of all the dissimilarities separating these two people, had a profound influence on his son. They both had powerful personalities. August, the father, had a stormy temperament that tended to be unbalanced, being given either to moods of contagious euphoria or to sudden outbursts of anger, or occasional depressions. In temperament the son, Werner, more closely resembled the mother. He was well-balanced, a happy boy, at peace with himself. He avoided confrontations; if his schoolmates started to fight, he went off in a different direction, so as not to get drawn into their squabbling. An intense, elementary feeling, even need, for harmony was his source of *joie de vivre* and deep compassion. And even as a child this penchant for measure and harmony endowed him with a rare self-possession, a sure instinct for differentiating between happy, spirited play and foolhardy, exaggerated bragging, something he avoided wherever he could.

August Heisenberg, the father, had had an unusual career. He was sent to a school for higher education at the instigation of the parson, who had recognized the boy's remarkable abilities at an early age; he passed the *Abitur** and subsequently studied classical languages in Munich. Following his graduation he became a teacher in a secondary school in Würzburg, where his great pedagogical talent was recognized. Simultaneously with his school activities, he passed his *Habilitation,* and soon thereafter was offered a post at the University of Munich, where he occupied the only chair for Byzantine Studies existing in Germany at the time. He held this position for the rest of his life. His name is still known and treasured by scholars in his field today. Munich was August Heisenberg's chosen home: it is where he had studied, married; it is where he found nourishment for his manifold, especially musical, interests; it is where

---

* *Abitur:* the final examination taken at the completion of one's secondary education, necessary for admission to the University.

he settled down. He had a beautiful, strong voice, and, true to his temperamental spirit, he loved to sing arias and *Lieder*; it was one of his greatest joys that his son Werner could accompany him on the piano at such an early age.

As was normal in middle class circles during the period prior to the First World War, August Heisenberg was a patriot. Having ascended from the artisan into the middle class, he respected its ideals and its order. It went without saying that he thought of himself as a German and that, during the war, he would fulfill his "duty to the Fatherland," as it was called in those days. To him, this was a living reality. In spite of this nationalist component, he had a sure sense for the moral quality of politics, and as early as Hitler's first appearances on the political stage, in other words during the early 20's, he was warning his two sons about this "chaot and seducer." "Never have anything to do with this Hitler character!" he would tell them again and again even then. From a certain aspect, it was fortunate that he did not live to see the takeover of the government by Hitler. His temperament and his strong rejection of National Socialism would surely have put him in a position of utmost danger; he had none of his son's caution. August Heisenberg died in 1930 of typhoid fever he had contracted while on an educational trip to Saloniki.

It must have been around the time of the Kapp riots in Munich. The father of a Jewish family called Levi — they lived above the Heisenbergs in the house in the Hohenzollernstrasse — appeared at the door of Heisenberg's father one day and asked him to take a little sack of precious stones into his safekeeping, until the antisemitic storm subsided. Father Heisenberg was rather taken aback by this request and by the measure of trust it showed, and asked whether there were at least a listing of the valuables it contained. No, that would not be necessary, was the answer: "We have complete trust in you." Naturally, the little sack was returned to Mr. Levi later on, and no one ever saw its contents; its owner could take it along when he emigrated to start anew in a more tolerant country. This is recounted to demonstrate how, in the case of August Heisenberg, political views manifested themselves on a purely human level. The elder Levi knew quite well that father Heisenberg could not be blinded by a mere ideology; he could put his trust in him without concern. And this incorruptible humaneness was the most important inheritance to be passed on from father to son. He, as well, never offered an ideology, whether of the right or the left, a nourishing foundation. The intrinsic narrowness of all ideologies, the corresponding fanaticism, and the inhumanity resulting from it were deeply repugnant to him. His most telling characteristic was his inner freedom, his total independence from popular opinions. Essentially, he trusted and felt responsible only to his own critical judgment. "But he wanted to make sure of it himself!" This was one of the epitaphs given him by the family, and it was ever so characteristic of him. Basically, it was also one of the secrets of his scientific success; a speech he once gave to the recipients of the Humboldt scholarship makes this clear: "Whoever has dedicated himself to science has, indeed, already made the inner decision never to accept a way of thinking without review and criticism; rather, he must doubt and doubt again, must examine, and must remain open to all opposing arguments." This inner freedom and independence was a character trait that appeared in him at a very early age, and it opened the possibility for him to act without prejudice and in a humane way in difficult situations.

From his earliest youth, young Heisenberg was extraordinarily involved with mathematical questions: this is generally known. Initially, it was probably more of a "sporting" interest in being able to understand and solve mathematical problems. But very soon, philosophical questions that are connected with mathematics entered into his consciousness. He describes this very vividly in the first chapter of his autobiography. Along with the theories of cognition and of relativity it was, above all, Weyl's book *Space, Time, Matter* that captivated him to such an extent that Professor Lindemann, the mathematician at the University of Munich, whose advice he had sought after completing the *Abitur*, rejected him for "already being spoiled for mathematics." Professor Sommerfeld then productively channeled this combination of interests toward physics.

This strong preoccupation with the general questions of natural science at this early stage certainly dulled his awareness of political events, and perhaps prevented a growing political consciousness from materializing. Basically, he was an unpolitical person. His friend Carl Friedrich von Weizsäcker formulated this very precisely in a memorial address: "He was," as he put it, "first of all a spontaneous person, following that a scientific genius, next an artist close to the creative gift, and only in the last instance, out of a feeling of duty, a 'homo politicus'."

In his early years, Heisenberg substantially adopted the political attitudes of his father. This explains why he let himself be induced almost unthinkingly into the ranks of the government troops during the upheavals under the revolutionary councils. At the time he was 17 years old. In his memory this period represented a strange mixture of exciting and even dangerous "war games" and agitating, brief glimpses into a rough and terrifying reality. When he looked back, the aspect of the "soldier game" exhausted itself in the "conquest" of bicycles or typewriters from "red" administrative headquarters and similar exploits. He rarely talked about the bad experiences. He was present when a youngster barely older than himself accidentally shot himself while cleaning his gun and died screaming in agony. And once he had to guard an older man for a whole night. The man had been taken prisoner for being a "red" and was supposed to be sentenced the next day. The outcome of these trials was common knowledge. During the night he listened to the life story of this man and, in the end, was convinced of his innocence. Here, he felt, he was confronted with a true, deeply moving human tragedy for the first time, and he knew that he bore some of the responsibility for the man's fate. It is typical of him that he did everything in his power the next morning to free the man. He succeeded in getting as far as the captain, and he convinced him that he had to let the man go. Thus he saved his life. Here, for the first time, we find the trait that we had already encountered in the father: for Heisenberg political action took place in that inner free zone, and political convictions only really manifested themselves on a humane level. This also explains his following statement that became so well known later on: "The value of a policy is not recognizable by the ends it strives for, it is recognizable by its means." Basically, this occurrence during the revolutionary period in Munich already bears the imprint and reflection of his entire political attitude.

So as to understand his later political stance, it is necessary to be aware of still another side of his experience as a youth that took on a fundamental importance in his decision to stay in Germany: it is his encounter with the youth movement. He had the most joyous and profound experiences of his life amidst its untroubled freedom, savored together with happy people who shared his convictions.

The youth movement introduced two new dimensions into Heisenberg's life: nature and friendship. He tells us himself how it all started in the first chapter of his book *Physics and Beyond:* "Then, one afternoon, it happened that I was addressed by an unknown boy on the Leopoldstrasse: 'Do you already know that all the young people will be meeting at the Prunn castle next week? We all want to go along, and you should come too. We want to figure out for ourselves how things are to go on now.' His voice had an urgency I had never heard before. And so I decided to go to Prunn castle." So much in his own words. And then

Heisenberg describes this meeting in the romantic castle situated high above the Altmühl Valley, where the deepest problems of the world were discussed in the manner of those days: with a pathos that is best avoided nowaways, and with which Heisenberg himself did not agree. Nevertheless, he was touched and was caught up in the intensity of the moment, uniting youthful will and strength, nature and music in such a singular manner. Beyond doubt, for Heisenberg the experience of the youth movement was interwoven with much romanticism. Life in the forest by the campfire, with music and the singing of old, beautiful songs was bound up with a direct, powerfully felt experience of nature: they hiked through the open country, breathed the freshness of the morning, slept under the open stars or in the hayloft of a farmer, showing their gratitude by helping in the stalls or with the hay. It was an escape from the parental home, not much different from what parts of today's youth long for: a simple life, undistorted by civilization, free from the shackles of dead conventions. Then, too, it was an escape from the standards of the narrow middle-class life, and it is understandable that Heisenberg's father reacted just as critically and anxiously to it as contemporary fathers do to their children's rejection of the norms.

For many years Heisenberg was the leader of a troop of Boy Scouts. Someone who had been a member of this troop later told me how this came about. He described how two pairs of brothers, together with some friends, were discussing the possibility of organizing a hiking club while they were standing in the courtyard of the Max-Gymnasium. They deliberated who should be the leader of the group. "Do you see that fellow over there?" one of them said, "he is the best, you can rely on him." And they quickly agreed to ask Werner to be their leader. They had not been mistaken; the decision resulted in lifetime friendships, friendships that outlasted all of the coming turbulent times, even though they were intensively tested by experience. Only one single boy of this group later joined up with the National Socialists, a very small percentage indeed!

There is no doubt that the youth movement had its political aspects, as well. The youth movement of northern Germany, especially in Berlin, was highly influenced by "first-class communist" tendencies, as we said in those days, and international solidarity played an important role in it. The boy scouts in the south, the group of which Heisenberg was a member, also had their left wing, but the faction displaying a more national, German race-oriented character, was highly predominant. Heisenberg was especially skeptical of this political side of the youth movement. In his opinion, there was no room for political agitation in the movement. He believed that the best possibility of educating people to be spirited and without prejudice was in an atmosphere unhampered by ideology; then they would be able to make mature political judgments and decisions.

Once he, too, had a confrontation with the national, race-oriented faction of the youth movement, in which the boys had to prove their loyalty with rigorous tests of their courage, had to underline it by reciting Germanic oaths of allegiance, and where they dreamed about a newly strengthened great German empire. After a few heated arguments, Heisenberg and his group immediately distanced themselves from this, and he stalked out of the middle of the meeting. [I believe it took place in the *Fichtel* mountains] under protest with all of his boys. From then on he always stayed well away from these groups. Indeed, in his student years, Heisenberg turned his attention more toward the socialist circles of the youth movement, who saw their most pressing duties in the propagation of new pedagogic impulses.

After the first world war a major movement had come into existence that had set itself the goal of encouraging the uneducated population, the workers, to participate in the cultural riches of their country. Today it is almost impossible to imagine the revolutionary character of his movement.

Young people from wealthier middle-class families were becoming elementary school teachers; this marked a courageous rejection of all traditional values and caused a great uproar at the time. These young idealists wanted to introduce music and learning, art and literature into the elementary schools; not as goods for consumption, but as a creative reality, so that the stunted intellectual abilities specifically of the lowest classes could free themselves to actively expand and develop. Thereby the hope was to guide the working class to a greater political self-awareness, to independence, and a capacity to make its own judgements and to reduce the deep gap between the ranks of society. The idea of adult education and the introduction of evening schools for continued education are among the most striking pedagogic experiments of the time. Heisenberg was enthusiastic about these ideas; they corresponded exactly to his own persuasions, and he immediately put himself at the disposal of the school for continued adult education in Munich. He gave courses in astronomy for workers, and at night he took his interested audience on excursions outside of the city and explained the firmament and its fascinating mysteries to them. On another occasion — later he was always to think of this as somewhat pretentious — he tried to initiate the workers into the beauty of the world of Mozart's operas together with a music student who was trained in singing. Since all this took place with carefree enthusiasm and he was already an outsanding pianist at the time, obviously even this course was a success. Heisenberg was deeply impressed by the openness and commitment of his students, who, for the most part, were people who had come back from the war starving for spiritual nourishment. For the remainder of his life he felt a special responsibility toward them. They belonged to the human sphere in which his political decisions and actions were rooted and in which they ripened.

1 August Heisenberg, the father, professor of Byzantine culture at Munich

2 The Heisenberg family (around 1906) on a fieldtrip in the vicinity of Würzburg

3 The father with his two sons, Erwin, the elder, and Werner (1914)

4 Arnold Sommerfeld, Heisenberg's teacher in theoretical physics in Munich

5 Werner Heisenberg 1927 (photograph by Fritz Hund)

6 Max Born and Heisenberg 1947

7 Niels Bohr

8 Heisenberg at his inaugural lecture in Leipzig (1927)

9 The conference in Copenhagen in 1930; first row (l. to r.): Oscar
   Klein, Niels Bohr, Werner Heisenberg, Wolfgang Pauli, George
   Gamov, Lev Landau, Hans Kramers

chapter two

FIRST CONFLICTS WITH POLITICS

The first time Heisenberg was confronted with political problems that affected him directly was in 1922. In the summer of that year his teacher Sommerfeld made it possible for him to listen in on a lecture by Einstein given at a convention of German natural scientists and doctors in Leipzig. As he entered the hall where Einstein was to speak, a pamphlet was pushed into his hand. It polemicized against the "Jew" Einstein in an inflammatory manner, stating that he was warping the laws of simple, classical physics with his consumptive, alien speculations. When Heisenberg learned that the originator of this pamphlet was Philipp Lenard, a world-famous physicist who had been awarded the Nobel Prize for his earlier research, he felt a world collapse. He had been living in the faith that science was determined solely by its striving for truth and insight and that it was impervious to political machinations. He was deeply shaken by the realization that this was an attempt to combat scientific truth with political means. In the fourth chapter of his autobiography, he describes his reaction as being very emotional, more confusing and depressing than conducive of action. He was so disturbed that he failed to notice that the man whose lecture he had heard was not Einstein at all [Einstein had cancelled because of the antisemitic agitation], it was von Laue; at the time Heisenberg did not know either of them yet, and he returned to Munich the same night, deeply alarmed and depressed by what he had experienced, without even attempting to contact the speaker in Leipzig. Since his belongings had been stolen from the cheap quarters he had moved into in Leipzig, as well, Heisenberg's first step was to hire himself out as a woodcutter in the Forstenrieder Park, so as to earn the money he needed to replace his possessions, but also to regain control of his emotions, and to clarify his thoughts on what he had experienced. From that point on he considered a political movement utilizing this type of measure to be basically immoral and incapable of having the slightest power of convincing him; this decision was final. At that time, however, there was no active involvement yet, no personal political *engagement*.

This political inactivity must be seen against the backdrop of his scientific creativity and the dramatic developments that physics was undergoing at the time. The field was, as has been described so often, in a state of extreme instability, in which countless numbers of scientific problems were stirring the minds of a young guard of scientific talents. The deeper the insights into the world of the smallest building blocks of matter penetrated, the more urgently the seemingly incomprehensible and insoluble contradictions to the laws of classical physics moved into the forefront. For someone with the talent of Heisenberg this state of the science held the greatest fascination. His teacher, Arnold Sommerfeld, had known how to promote this fascination. From the beginning he had constantly confronted his young, gifted pupil with the central problems of the most advanced atomic physics. And now, in June of 1922, he sent him to Göttingen, where the Danish physicist Niels Bohr was to hold a series of lectures on his model of the atom and the problems connected with it. With the large circle of young, talented physicists from all over the world that Niels Bohr had gathered around himself, he was considered to be the uncontested center of pioneering work in the field of modern theoretical atomic physics. And now Heisenberg, who was barely 21 years old, had the opportunity to hear and experience this world-renowned man who was the master of his discipline.

Heisenberg was immediately deeply impressed by Bohr's personality. Never had he encountered a man whose words revealed so much inner consistency and depth. In "Reminiscences of Niels Bohr," a lecture he gave in Copenhagen upon receiving the Niels Bohr Medal in 1964, he recalled his first meeting as follows: "Full of youthful vigor, but still somewhat uneasy and shy, his head inclined slightly to one side, the Danish physicist stood on the bright rostrum of the auditorium, while the broad light of the Göttingen summer streamed through the wide open windows. His sentences seemed rather hesitant and soft, but behind every one of the carefully chosen words you could feel a long chain of thoughts disappearing somewhere in the background of what, to me, was a very exciting philosophical viewpoint." Physicists later named these famous lectures by Niels Bohr the "Bohr-Festival"; they were the prelude to a great and stormy development in physics, during the course of which quantum physics was created and Heisenberg developed his uncertainty principle.

In spite of his youth, Heisenberg ventured a critical observation during Bohr's second or third lecture. Bohr was immediately attentive. He instantly sensed an unusually clear penetration of these difficult problems in the objections of the young man, and was surprised by the confidence with which he dealt with questions of detail. And he quickly recognized the great creative ability in him. After the lecture he invited Heisenberg to take a walk with him on the Göttingen Hain mountain so that they could continue their discussion; actually, though, he wanted to get to know this interesting young man more closely. Heisenberg wrote about it later: "This discussion that led us back and forth through the wooden heights of the Hain mountain was the first intensive exchange about atomic theory I can recall; it certainly played a decisive role in the development of my future life."

Since this encounter, it was Heisenberg's most fervent

wish to travel to Copenhagen to get together with Niels Bohr and to join the group of international physicists who, inspired by the intellectual richness and depth of their teacher, were discussing and working on the exciting central problems of atomic physics. But first he had to complete his education; in other words, he had to finish his doctorate in Munich and subsequently obtain a position at a university. He received his doctorate one year later, though by no means as brilliantly as everyone had expected him to. For, completely absorbed in the problems of atomic physics and their theoretical solutions, he aroused the anger of the experimental physicist Wilhelm Wien, his second examiner, who considered Heisenberg's knowledge of experimental physics to be inadequate. "It would seem he wanted to flunk me!" Heisenberg would say if the conversation came around to this topic. "I was much too committed to theoretical atomic physics, and the experiments we were performing in Wien's courses struck me as being a sheer waste of time." Precisely this attitude, however, provoked the professor, who was skeptical about the many "geniuses" — as they were often jokingly called — in Sommerfeld's seminar. Even so, this failure disheartened Heisenberg; his father wrote him, saying that his academic career appeared to be over with once and for all. Heisenberg, however, could not quite believe it, and so, for two weeks during the Easter vacation he traveled to his new mentor Niels Bohr in Copenhagen. This period marked the beginning of a new friendship, to which Heisenberg felt committed — indeed, indissolubly bound — for the rest of his life.

During these days with Niels Bohr, Heisenberg also received his first instructions in politics. For three days Bohr roamed through his beloved Denmark with his young guest. He showed him the old historic locations and acquainted him with the history of the country and its old sagas, thereby clarifying the spiritual structures that were still in

effect in Scandinavia: the spirit of freedom and independence, treasured more highly than happiness and life and
the power of a great empire. For Heisenberg, to whom the
concept of a great empire that the individual had to serve
and, if necessary, sacrifice his life for, had been an exalted
ethical demand, this proud song of freedom and independence now being recited to him by Niels Bohr disclosed
a whole new world. He understood that this spirit had given
birth to the idea of democracy, an idea that was so much
more a reality in Scandinavia and in England than in what
we experienced in Germany. I myself was given an intuition of how impressed he was by this some time later. On
one of our first meetings he told me about it, about his
roaming through beautiful, free Denmark with Niels Bohr,
about this totally different mentality, the democratic conviction that was the unquestioned foundation of the culture
there. Ever since, Heisenberg felt unflaggingly committed
to this spirit, and he carried an image in himself, in which
the individual unflinchingly meets his fate, be it fortunate
or adverse, in order to stay true to the call of his own inner
spirit. He lived accordingly, with the consistency and steadiness of purpose so characteristic of him. He, too, felt that
he had set out to embrace and solve certain problems. He
had to act in consonance with this.

The days in Denmark passed all too quickly and
Heisenberg had to return to Germany. He still needed to
obtain a lectureship at a university — only then would he
be completely independent. For this purpose, he had accepted the position as an assistant with Max Born in Göttingen. And on July 24, 1924, one year after his doctoral
examination, he was recognized as an academic lecturer.
Now the path was free: for the winter semester of 1924-25
he received a Rockefeller stipend for his first academic
sojourn in Copenhagen.

But the first meeting with the Copenhagen Circle, with
which Heisenberg was later to feel so closely connected,

was by no means as simple and unproblematic as he had expected. In his own words: "After a few days, my first introduction into the circle of young people surrounding Bohr at the time resulted in a deep depression. These young physicists from the most disparate countries of the world were far superior to me. Most of them were fluent in several foreign languages, while I could not express myself reasonably in a single one; they knew their way about in the world, played musical instruments with consummate skill and, above all, understood modern physics far better than I did." In reality, the situation was a great challenge to Heisenberg. First, he needed to learn languages. He had learned no English in the humanist *Gymnasium*; true, he could understand scientific treatises, but he was completely incapable of expressing himself in English. But he was in Copenhagen and was supposed to give lectures there — so he had to learn Dishish, as well. From his landlady, the friendly, elderly Mrs.Måar, who soon loved her young tenant like a son, he learned Danish and English regularly every day for several hours. Mrs.Måar was a good teacher, and she supported him wherever she could. In a few weeks he had caught up with his fellow students and could express himself fluently in English as well as in Danish. That was one side of his life in Copenhagen: on the other, he threw himself into his scientific work with the same energy and intensity. During this period he must have possessed unbelievable vigor, getting by on just a few hours of sleep.

Following this first summer in Copenhagen, Heisenberg rotated back and forth between Copenhagen and Göttingen. "I learned physics, along with a dash of optimism, from Sommerfeld; from Max Born, mathematics; and Niels Bohr introduced me to the philosophical background of scientific problems," he said of this learning period of his life. But it was not only a learning period; above all, it was a time of great creative force in scientific matters. Between 1922 and 1927 the record displays 27 treatises, most of which are of basic scientific importance. The inner dedication necessary to achieve this must have been of the utmost intensity. Heisenberg was completely absorbed by the scientific problems confronting him; they filled out his days and nights. The discussions with the other young physicists, with the sharp and critical Wolfgang Pauli, and, above all, with Niels Bohr himself, demanded all his energies. He often felt that he could not go on, that the strain and the exhaustion were becoming too much. At times like these he would retreat into the mountains; skiing and hiking were his irreplaceable means of relaxation. Once he travelled to Norway. In a book by Nansen he had read that to become a true man you had to confront the solitude of nature at least once. This he did. He hiked through the mountains in the north of Norway, through the snow-covered landscape void of human beings, making his way from one hut to the next [they were frequently separated by a journey of several days] with the aid of a compass and maps alone; once he broke through the snow covering a stream. In the summer, when he was plagued by hay fever, he retreated to Helgoland. Thus, he replenished his strength again and again, his head cleared, and he advanced further and deeper into completely new fields of thought, into a virgin country, in which he believed to recognize connections that would, of necessity, change the way man thought. As proof of his emotions, we may, at this point, cite the words he himself wrote, when, during the course of a night on Helgoland, he broke through to a clear vision of his problem: "In the first moment, I was deeply shaken. I had the sensation of looking through the surface of atomic phenomena at a distant foundation of a strange inner beauty, and the thought that I was to probe this wealth of mathematical structures that nature had spread out before me down below almost made me dizzy. I was far too agitated to sleep. And so, with dawn beginning to break, I left the house and headed for the southern tip of the island, to where a solitary tower of rocks jutting out into the sea had always kindled my desire to climb it. I reached the top without too much difficulty, and waited for the sun to rise."

In this time of intense concentration and emotional involvement in science there was no room for politics. By the intensity of his life Heisenberg was so isolated from political reality, growing ever more menacing in Germany, that he could only vaguely perceive it on the periphery. Even when he was offered a position at a university this initially did not change.

Heisenberg turned down his first offer in favor of the fruitful collaboration with Niels Bohr. It came from Leipzig in 1926. But as the people there were eager to engage the young, talented scientist for their university, they did not give up. The following year they repeated their offer and proffered the Professorship for Theoretical Physics. Simultaneously, he received an offer from the University of Zürich. Now he could no longer decline. After thinking about it for a while, Heisenberg decided in favor of Leipzig. Later, I once asked him why he had decided on Leipzig, and not the much more beautiful city of Zürich. His answer was quite spontaneous: "I preferred to stay in Germany." For him, Germany was the country where he had spent a fulfilled and lively youth; it was where he felt he belonged. The shadows of politics had not yet reached him at the time.

Thus Heisenberg went to Leipzig in the fall of 1927. Now his task was to create something similar to what he had experienced so uniquely in Copenhagen: a lively scientific discussion, combined with intensive human encounters, in a circle of talented young men from all over the world. In fact, in the five years preceding the seizure of power by the Nazis, a rich and animated scientific life did develop. Many gifted young men from all over the world came to the professor who had become so famous so soon, for he always discussed and seized upon the most contemporary problems. His teaching method must have been free and relaxed, the tone set between the professor and his students and colleagues, who were often older than himself, pleasant and humorous. During the summer, the seminar frequently took place in a nearby swimming pool; during the winter they played ping pong and met for afternoon tea, to which the professor donated the cake. And in March, when the sun was already gathering force, he and his most trusted and qualified retinue would travel to the Untersberger *Alm** on the Brünnstein mountain, where he had the use of a little cottage that years earlier, together with his group of Boy Scouts, he had restored from a condition of complete decay. There they skied and talked science intensively, laughed happily and philosophized. During this period names like Teller, Bethe, Bloch, Weisskopf, the young Weizsäcker, and many more can be found in his seminar, which quickly gained international recognition and joined the ranks of the other great scientific centers, such as Göttingen or Copenhagen.

In this animated atmosphere, sustained by the "international family of physicists," the threatening reality of politics was still delegated to the background. After his earlier experiences, Heisenberg found politics to be something sinister and alarming. And apparently he calmed his conscience by trying to meet the political challenges through his science and, together with the many people from all over the world who were gathered in Leipzig, the common search for the truth, by working and striving for the reconciliation of all peoples, for humanism, and for the abolishment of racial and ideological prejudices. This was his hope; and Niels Bohr was his model and his great teacher — not only in physics.

After the Nazi takeover in January, 1933, this all changed, not suddenly, but step by step, and politics became painfully tangible for Heisenberg as well. Slowly, the ranks in

---

*Alm: a small dwelling on the slope of a mountain where cows are taken for summer pasture.

his seminar began to thin out. One after the other those who had a chance of finding a position abroad left. But he himself was also drawn into conflicts with the National Socialist organization; he was a thorn in their side, for he refused to become a member of even one of their groups — indeed, he refused to cooperate with them. He had been requested to participate in a proclamation for Hitler and to sign a telegram, in which he would have declared his allegiance to the *Führer* — he refused to sign. He declined to participate in a gathering of teachers for the purpose of voting "yes" to Germany's withdrawal from the League of Nations; this became the subject of the unpleasant and dangerous situation he was then exposed to.

In recognition of his achievements in 1932, Heisenberg had been awarded the Nobel Prize in 1933. This put him in the spotlight of political attention. In November, after the press had circulated the news that the prize had been awarded, a friend informed him that a band of National Socialist thugs intended to break up his momentous lecture on the following morning; the ovations Heisenberg's students planned were to be turned into a demonstration against him. It was known that this type of action could be dangerous. Not infrequently they escalated into veritable inside battles, occasionally claiming lives. What was to be done? The friend — let us call him W.W. — who had warned Heisenberg, and who himself wore the uniform of the SA, offered to "take care" of the matter, as he put it.

The next morning the auditorium was overflowing. An uneasy, tense mood prevailed. Many had come to celebrate the young winner of the Nobel Prize, but now there were whispers about a demonstration against him due to his lack of solidarity. Amidst this atmosphere of insecurity, W.W. climbed up on a bench and, having attracted the people's attention, said roughly the following: he personally knew Heisenberg; he was, so he had always thought, a decent, patriotic fellow, but his refusal to join the teacher's gathering was beyond comprehension. If it were true that he had done this only out of consideration for his Jewish colleagues abroad — and that would indeed be scandalous — then he was finished with Heisenberg. But before this had been cleared up there could be no demonstration against Heisenberg. He had just received a telegram from the leadership of the party that, for the time being, there was to be no action against Heisenberg; this was a strict order, and violators would have to reckon with expulsion from their organization. The only remaining possibility would be to receive Heisenberg with icy silence. However, those who wished to manifest their displeasure even more strongly should now demonstratively leave the auditorium with him — and W.W. left the room with a small group of radicals. While they were marching out the upper door, Heisenberg entered by the lower one. He was greeted by a deathly silence. Depressed, he walked to the rostrum. Suddenly the first stomping of feet could be heard, it gathered with the force of a storm, and enthusiasm broke forth. After everything had calmed down again, the lecture could now take its course without further disturbance.*

There can be no doubt that all this was an extremely shady, prearranged affair, totally inscrutable to the uninitiated; so things had already reached the point where methods of this kind had to be used to prevent even worse things from happening.

How could all this come about? One will perhaps wonder how the students could be manipulated so easily and quickly. The fact that Heisenberg had a friend in the SA who utilized such shady methods will also seem strange to some seeing it from the vantage point of today. This can only be understood by knowing what it really was like at the time. A large percentage of the population as well as of the students, it is true, had decided to back Hitler; following the humiliations of the post-war period they wanted

---

*I received these details from Professor Rozental of Copenhagen.

a renewal of Germany's strength, and they hoped that a resolute leadership would be able to put an end to the confusion and miseries of the time. But the hate-filled and hypertrophied Party ideology had by no means permeated everyone, and only a few felt any solidarity with the use of terror to suppress differing opinions. Most people were insecure and frightened, and were not really sure what things were leading to, or what to believe. Many thought that National Socialism could still take a turn for the better, and they lulled themselves with the hope that the horrors they heard about were excesses of the initial period. And they felt called upon to add themselves to the equation by joining either the party or the SA. At the time none of us had any experience with regimes of terror, and many a person nourished the illusion that things could perhaps be made better by working from the inside out. But demoralization had made rapid progress; already, many no longer dared to operate in the open. W.W. employed lies and cunning to protect Heisenberg. There was no telegram from the Party leadership, let alone an order to spare him. Indeed, this whole game of lies is a sad and shameful example of how it was frequently possible to protect one's self from harm only through the use of sordid means. And this at this early stage.

Heisenberg had known next to nothing about the manipulations that W.W. had concocted to protect him. Upon entering the auditorium he felt the icy, eerie silence [he told me about this later], and he was relieved when it turned into applause and concord; W.W. had taken care of this, as well. Basically, W.W. was a decent person; to me he did seem to be somewhat dull and perhaps weak, and intellectually vague. But Heisenberg was firmly convinced that he would never let himself be misused for wicked purposes. In this dangerous situation he had placed his trust in W.W., and was always grateful to him for having spared him a most unpleasant experience. Later, W.W. realized that he was on the wrong path; and in deep despair over the events he had to share responsibility in, he quite consciously sought his death in the war. His farewell letter to his family is a deeply affecting document of those confused times.

Soon it could no longer be denied that, rather than getting better, conditions were worsening with a terrifying consistency. In 1934, in an address given at the convention of natural scientists in Hannover, Heisenberg unequivocally and clearly backed Einstein and his teachings. This resulted in an attempt to isolate and destroy him by the National Socialist element among the physicists, culminating three years later in the abuses of the "Black Corps." In the universities, as well, the National Socialists were increasingly gaining the upper hand. In Leipzig, Heisenberg fought against the dismissal of deserving, highly distinguished colleagues, and he fought with verve; now we know that this endeavor was completely futile, at the time we did not know it. At the time we were beginning to find out. Not only our colleagues became victims of the National Socialist race-mania, but so did our friends. Our closest friend, the attorney Erwin Jakobi, a great violinist with a versatile and most charming personality, to whom we were bound for life through music and a close personal friendship, was dismissed from his position as a professor. All we could do was to support and encourage him with unbroken amity. Heisenberg's best people left him; indeed, he himself had to advise them to depart Germany, something he found very difficult to do.

Under these circumstances, Heisenberg felt his duty to be in more effective actions. Together with three of his colleagues, the physicists Fritz Hund, the mathematician Bartel Leendert van der Waerden, and Carl-Friedrich Bonhoeffer, professor of Physical Chemistry and a brother of the theologian Dietrich Bonhoeffer, who was later executed by the Nazis, possibilities of steps that were needed to signal their unwillingness to go along with what was happening were discussed, a signal meaning: "So far, and no farther!" They thought about collectively resigning from their teaching positions — certainly a spectacular demonstration. Heisenberg was not quite sure that such a move would make any sense, would have any effect. In this situation, they decided that Heisenberg should travel to Berlin to see Max Planck to debate their plan with him. Max Planck was a generation older than Heisenberg and his friends, he was experienced and upright in his convictions — he was above any suspicion. But Max Planck advised them against it. He told them that nothing would change if they left. The radio and the newspapers would report that the regime had no longer been able to support the four professors, and that they had to be dismissed from their positions due to their subversive agitation hostile to the people. He said that an avalanche, once in motion, could no longer be influenced. That, he said, was the reality of the situation. "Hold out until it has passed, form 'islands of stability' and salvage things of value from the catastrophe." That was the quintessence of his advice to Heisenberg.

Heisenberg was deeply discouraged on his return trip to Leipzig. He recognized that it was too late for an individual to affect the course of events, that fate would inexorably roll over his friends and himself. Nevertheless, he decided to follow Planck's advice, as did the others. No doubt, the conversation with Max Planck was certainly an essential factor in his subsequent momentous decision not to emigrate from Germany, a decision made up of so many different components. To form "islands of stability" — that made sense to him and embedded itself in his mind.

chapter three

"GERMAN PHYSICS" AND THE SUCCESSION OF SOMMERFELD

Heisenberg's experience in Leipzig, where Philipp Lenard had so polemically accused Einstein of practicing subversive physics, was unfortunately only the prelude to a larger movement disruptively invading and confining Heisenberg's life, and finally kindling his political activity. In 1933, Einstein had left Germany for good and had finally settled in his new country, the United States. His departure from Germany had cleared the way for those elements in science who thought that general problems of the natural sciences could be solved on the level of an "anti-semitic spirit." Philipp Lenard was the best thinker of this movement. Since, following his great, earlier successes that had won him the Nobel Prize in 1905, he had not been willing to switch over to the modern international development of physics, it was logical for him as an outsider to declare his solidarity with the elements of the National Socialists that were doing battle with the abstractions of modern physics on racist and ideological grounds. The main force of his attack was centered on Einstein and his theory of relativity. To Lenard, Einstein was the prototype of the "decadent Jewish spirit," who was betraying the simple and clear principles of classical physics. Lenard was the "inventor" of "German physics" [also called "aryan" physics], which, though it could not solve the problems of modern physics, tried to suppress and discredit it as being "alien" by evoking the principles of classical physics.

Lenard's great ally was Johannes Stark, like himself a respected physicist, who had won the Nobel Prize for the discovery of the "Stark-Effect" carrying his name. In the battle against modern physics he was the strongest driving force and a truly dangerous, unscrupulous adversary. Unfortunately, there was also a fairly significant faction among the younger scientists, who joined in with and supported the proclamation of "aryan physics." Since Lenard and Stark had official government backing, they attracted not only those opportunistically considering their own advantage, but also those incapable of following the difficult trains of thought demanded of them by modern physics. And finally, there was the group of uncertain and confused people who did not know what to think. This was the assemblage with which Lenard and Stark were operating. In this way, "aryan physics" could gain influence on the development of the sciences in Germany in spite of its lack of scientific qualifications, an influence that appeared thoroughly ominous to those who felt responsibility toward the course of this development.

The ranks of physicists still teaching modern, true physics had thinned considerably. Many of the leading personalities had left Germany: besides Einstein, Max Born, James Franck, Heisenberg's outstanding assistant Felix Bloch, Hans Bethe, as well as many others. Lenard wrote triumphantly: "Of its own free will, the alien spirit is already leaving the universities, indeed, the country!" Although he was deeply troubled by this development, on his return trip from Berlin Heisenberg had made the decision to follow Planck's advice and to remain in Germany. In spite of everything, he was determined to continue to instruct his students in what he had recognized as the truth and in what he had learned from his own great teacher Niels Bohr. He was not beaten that easily and was not willing to make his retreat. Later he wrote about it: "I decided to stay, probably due to the feeling that the fate of Germany was sealed, unless it proved to be possible to eliminate the absurd and criminal features of National Socialism from within. At least with regard to physics, I did not believe I could relinquish the duty of doing this; Bohr, as well, seemed to approve of this standpoint." [This was written during his detention in the fall of 1945.] Thus he took up the fight with the specter of national hubris manifesting itself so clearly in "aryan" physics.

The retirement of Sommerfeld in Munich in May of 1935, at the age of 67, caused an unexpected escalation of this battle. Sommerfeld had been an exceptionally upright man, and had never attempted to hide the fact that he backed Einstein, Niels Bohr, and modern physics. Moreover, his professorial chair in Munich was held in very high esteem, as Sommerfeld was a highly talented, committed teacher who had trained a whole generation of capable physicists. Thus the filling of this position was an issue of central importance.

As early as 1935, Sommerfeld had asked Heisenberg whether he wanted to come to Munich if he were offered the position. Sommerfeld himself very much wanted to see Heisenberg as his successor. This offer delighted Heisenberg and was, being his most ardent desire, a great temptation. I found this passage in a letter he wrote to Sommerfeld after the war: "My heart...conjures up the image of the blue sky of the Bavarian foothills, the memory of my student years with you, and the unbelievable radiance of that earlier Munich...my heart reminds me of sailing on the Lake of Starnberg, of the powdery snow in front of the Untersberger Ski cottage, and the air in spring, when the warm wind, swooping in from the deserts of Africa, was blowing down off the mountains." He hardly ever waxed so lyrical, and only once, during his now famous speech given at the 800th centennial of Munich, did he again express his love for that city with such moving words. But, above and beyond this, Heisenberg felt that, in a certain sense, the chair of his teacher was, by right, more his than anyone else's. And after all, his summons to Munich would have been an immense victory over "German physics" — but precisely that proved to be the undoing of the plan.

Not only Sommerfeld wanted Heisenberg to be appointed; he also had the approbation of the members of the Munich faculty who were still upright. In a letter written to the Ministry by the appointment committee, consisting of the physicists Gerlach and Sommerfeld, the chemist Wieland, and the mathematician Caratheodory, dated July 13, 1935, we can read the following: "Of Sommerfeld's pupils, W. Heisenberg is the most famous and important... Heisenberg's versatile productivity is singular, as is the vital energy he transmits to his pupils."

But in the meantime, the Ministry was permeated and controlled by the Nazis, and the followers of the maxims of Lenard and Stark had already gained the upper hand. So the proposal of the appointment committee was confronted with intense resistance. There was no interest whatsoever in stirring up trouble by letting Sommerfeld's star pupil loose in their own back yard. Heisenberg had already made himself too unpopular. He had not joined the party and was a declared opponent of "German physics." Stark had already published a brochure, "National Socialism and Science," in 1934, in which he sharply attacked the exact natural sciences. As a counter-action, Heisenberg had delivered an address to the convention of natural scientists in Hannover in September of the same year, clearly formulating the viewpoint of true physics and defending Einstein and his teachings. In addition, this address was published in the *Zeitschrift für Naturwissenschaften.* Heisenberg's next move, together with Max Wien and Geiger, was a public declaration, in which the standpoint of Lenard and Stark was sharply rejected in a carefully prepared text. The declaration was signed by the majority of German physicists. The ministry could not tolerate all this, and Heisenberg was rejected.

But the battle raged on and was fought with increasing violence. On December 13, 1935, Stark gave an indignant speech during the celebration of the opening of the Philipp-Lenard-Institute in Heidelberg: "... Today, Einstein has disappeared from Germany. But unfortunately, his German friends and supporters still have the opportunity of continuing their work in his spirit. His principal promoter, Planck, still heads the Kaiser Wilhelm Society, his interpreter and friend, Mr. von Laue, is still permitted to play the role of an adviser for physics in the Academy of Sciences in Berlin, and the theoretical formalist, Heisenberg, the spirit of Einstein's spirit, is even to be distinguished with a university appointment." An article written in the *Völkischer Beobachter* by a supporter of Stark, a student by the name of Menzel, carried the conflict out of the scientific sphere into the realm of the broad public. Heisenberg was not sure whether he should respond or not; after all, the *Völkischer Beobachter* was the official mouthpiece for voicing attacks by the Party. Should he, dare he, get that involved with the Nazis? But finally he decided he had to speak out precisely on account of the public. On February 28, 1936, his article, written in an intentionally measured tone without a trace of polemics, appeared in the *Völkischer Beobachter.* In it he describes the cognitive value of theoretical physics, finally referring to Kant by designating theoretical physics as the continuation of the great philosophical tradition "initiated by Kant, using the theory of cognition to investigate the foundations of natural science." But Heisenberg had underestimated his opponent. Stark's response was directly appended to his own article. It begins with the sentence: "In the interest of clarification it is necessary that the preceding article by Heisenberg receive an immediate emendation." Initially, Stark tried to continue the measured tone used by Heisenberg, but the attempt quickly failed. Stark then placed the full blame of the controversy on the long-smoldering antagonism between experimental physics versus theory, claiming experimental physics as the true physics, embodying the old classical values of observation and recognition. All great discoveries had been made by experimental physics. Theory, on the other hand, is nothing other than "a method, a calculated and representational expedient...an aberration of the Jewish mind, that must no longer be allowed to have the decisive influence it has enjoyed up to now..." This turn of events was dangerous. Here was a real threat, directed not only against Heisenberg, but against theoretical physics as a whole, and thus against all theoreticians. Once again, Heisenberg felt called upon to continue the battle.

In spite of these exhausting controversies that never offered any security as to what would happen next, whether the so-called "truth" might not be decreed by the government one day [a not infrequent occurrence, in fact], for a while the situation in Munich seemed to be developing in Heisenberg's favor. A new chancellor had been assigned to the University of Munich, who, though he had the "right" credentials — he was a member of the party and faithful to its line — was nevertheless concerned with using his position to restore some of the former dignity and scientific integrity to the university. Thus he was interested in having Heisenberg come to Munich as Sommerfeld's successor to occupy the chair that had been vacant for so long. To this end he exerted his whole influence with the involved authorities. So, in the spring of 1937, it looked as though Heisenberg would soon be assuming his new position in Munich. Kölbl, the chancellor, had asked Heisenberg to come to Munich in the summer semester of 1937 as a deputy to start giving lectures and seminars. He had also received assurance from the Ministry that the appointment would then indeed be confirmed. Heisenberg had agreed and had already purchased a lovely house in the Isartal he hoped to move into during the summer. But things were to turn out quite differently.

At the end of January 1937, we had met at a musical gathering in the home of the Mittelstaedts. This evening decisively changed our lives. We both felt that we had encountered "our fate." Neither of us knew what this fate was to be, but that we would emerge from it as changed people was clear to both of us on that first evening. The friendly nudge coming from our solicitous hostess at the end of the evening: "Mr. Heisenberg, would you accompany Miss Schumacher home?" was hardly necessary. We were engaged ten days later, and planned to marry in April.

Due to this, Heisenberg asked Munich for a postponement of one semester. We both traveled to Munich in mid July. Heisenberg had cut short the semester in Leipzig, since I was not well and the doctor had urgently recommended a change of air. We took our baggage to the apartment of Heisenberg's mother in Hohenzollernstrasse, and, after exchanging greetings, Heisenberg immediately phoned the chancellor for news about his appointment. Kölbl was short on the phone, saying only: "Have you already seen the *Schwarze Korps*? There's a long article on you in it. Buy the paper immediately and read it. Then we can talk." The *Schwarze Korps* was the organ of the SS, and we were not in the habit of reading it. Heisenberg went to a newspaper stand immediately and bought the July 15 issue. The article in question was entitled: "White Jews in the Sciences." It was, indeed, a frontal attack on Heisenberg, and it was signed by Stark. The following passages are excerpts from this article: "How secure the 'White Jews' feel in their position is proven by the actions of the Professor for Theoretical Physics in Leipzig, Prof. Werner Heisenberg, who managed to smuggle an essay into an official party organ in 1936, in which he declared Einstein's theory of relativity to be 'the obvious basis for further research' and in which he describes 'one of the most pressing tasks of the scientific youth of Germany to be the continued development of theoretical conceptual systems.' Simultaneously, he attempted to garner the favor of those in authority through a vote of German physicists on the value of the theory, and to gag the critics of his actions. This representative of the Einsteinian 'spirit' became a professor in Leipzig, as the model pupil of Sommerfeld, in 1928 when he was 26 years old, in other words, at an age when he had hardly had the time to engage in thorough research..." And elsewhere: "Heisenberg was awarded the Nobel Prize in 1933 together with the Einstein-disciples Schrödinger and Dirac, a demonstration of the Jewish-influenced Nobel committee against National Socialist Germany, that corresponds to the 'honoring' of Ossietzky. Heisenberg returned his thanks by refusing to sign a proclamation of German Prize recipients for the *Führer* and Chancellor of the Reich in August of 1934..." Elsewhere Heisenberg was described as a "Jew lover" and a "Jewish pawn," followed by the statement: "His fame abroad is an inflated result of his collaboration with foreign Jews and Jew lovers."

Such an attack in a high official, militant organ of National Socialism was not without its dangers; no doubt, though, it meant the annulment of Heisenberg's appointment in Munich. That was the end of his succession of Sommerfeld; on this front the battle had been lost. From then on, his sole purpose was to impede the advance and dominion of the mendacious pseudoscientific teachings and their narrow philistine spirit in the universities.

In this context I would like to draw attention to a fact that has frequently been doubted, but is nonetheless an incontestable reality. Today we easily forget how difficult it was to obtain a truthful, realistic picture of events in those days, events transpiring behind a facade of muteness and fear and only occasionally illuminated by flashes of light so full of horrors they were nearly impossible to believe. We knew that we were living under a criminal regime. The difficulty of that time was that we had no experience with the possible extent of this criminality or with how far people would be willing to follow the criminal path. We were badly informed; not so much because we did not receive sufficient information [obviously, this was also the case]; but mainly, we were badly informed because we were incapable of believing the magnitude of the evil present in our own people. With this I am not trying to say that as Germans we thought ourselves better than other nations — specifically this type of hubris was, in our minds, connected with the philistine spirit of National Socialism. But we thought of ourselves as Europeans, and at the time Europeans still lived with the conviction of belonging to the civilized peoples of the earth, the peoples who had left unbridled barbarism behind, and who were dedicated to fair tolerance and humaneness. In school, and sometimes from our parents, we had learned that we were the nation of "poets and thinkers," and we were proud to believe it. Now we lacked the imagination to picture the organized crimes our people were capable of. If we did not have enough fantasy to recognize it, our consciousness refused to interpret the signs of the time. Today, knowing the whole terrible truth, it is barely possible for us, as well, to imagine how little we were capable of understanding back then. Even during the war people continued to deny the complete reality. I can still see my father standing in front of me. He was a man with a venerable and law-abiding outlook, who actually went into a rage when Heisenberg once showed him a report he had received from a colleague at the institute, who had been a witness to the first cynical mass executions of Jews in Poland. My father lost all self-control and started to shout at us: "So this is what it has come to, you believe things like this! This is what you get from listening to foreign broadcasts all the time! Germans cannot do things like this, it is impossible!" He was not a Nazi; he had prematurely retired from his position following the National Socialist takeover.

The most severe reproach we are subjected to by the other

countries of the western world is, indeed, the fact that a people with such a highly developed culture could regress into such a barbaric conduct. That, precisely, is what was so difficult for us in Germany to comprehend and understand to its full extent. We, too, could only grasp it in its totality when the concentration camps and prisons were opened up after the war, revealing images of unprecedented horrors, as ghastly for us as for the rest of the world, images that the whole world now envisions when it thinks of the Germany of the Nazi period.

Heisenberg was not a person to bury his head in the sand. Whenever possible, we listened to the English broadcasts. But Heisenberg loved the country of his childhood and youth; he did not believe that the picture that was now looming so appallingly was the true countenance of Germany. Within himself he carried the picture of another Germany for which he thought he had to persevere. He was determined to create an island of freedom and trust, just as Planck had advised him to. That he would fight for. Obviously, the opponent had now gained a victory, and Heisenberg was made to feel it: young students who had asked to study with him were withdrawing their applications, saying that it was too dangerous; colleagues who earlier had sought his company were now distancing themselves from him; and there was no lack of abuse from the ranks of the other lecturers and the student leaders. The following passage can be read in a letter to the *Reich* director of the National Socialist Party: "A concentration camp is doubtlessly a suitable location for Mr. Heisenberg...it is probably also time to prosecute him for betraying both the people and the race." The letter is signed by a secondary school teacher.

A conspicuous trait in Heisenberg's character was his stubbornness. It is said of the Westphalians that they are incredibly bullheaded, and Heisenberg was Westphalian! All of his paternal ancestors had been Westphalian farmers and artisans. The family members were quite aware of this, and we had a *bon mot* we occasionally teased him with; it can sometimes be seen as a bumper sticker affixed to the back of trucks or cars: "No use in honking, Westphalian driver!" The imperturbability with which he steered toward his goal and stayed his course was without comparison, just as it was certainly also a reason for his scientific success. It was his strength and, at times, his weakness. Coming from this deeply rooted characteristic, he had no option other than to continue the contest with all the means at his disposal. He was now confronted with the decision of either giving in to the pressure of the Nazis or of prevailing. He was in the position of a fencer who, his back to the wall, lances an attack to regain freedom of space, the space to continue his calling as a teacher. And all the while he knew he could only beat the enemy with their own weapons; so he had to take the risk that he took.

On July 21, 1937, one week after the publication of the article in the *Schwarze Korps*, Heisenberg wrote a letter to Himmler, the supreme leader of the SS in the *Reich*, and the highest authority responsible for the *Schwarze Korps*, demanding "effective protection from this kind of attack," and requesting that his "honor be restored." He sent a similar letter to Rust, the Secretary of Education, his topmost employer. Heisenberg was trying to initiate a disciplinary directive against himself, an inquiry in which the slanderous statements of the *Schwarze Korps* would be examined as to their actual verity.

At the time, I was very anxious and, in a certain sense, shocked by this move. While I had been studying in Freiburg, I had encountered several instances that had demonstrated that the justice of the National Socialists had nothing to do with right or wrong and that, once caught up in their doings, one was easily and quickly subjugated. Heisenberg had not discussed this move with me; probably he did not want to burden me with his decision, and I guess that he had an inkling that I would not have agreed. Even now, with a better understanding and more insight into his motives than I had then, I thought the stakes he was playing for too high, even if his success, and there is no denying it, did justify his actions.

There were moments throughout Heisenberg's life when he would take a risk that seemed to go against all reason. The most dramatic of these instances was probably at the end of the war when the allied troops had already occupied Hechingen. He was cycling to Urfeld to help his family through the last days of the war. During the trip he ran into an SS sentry who accused him, in a most unfriendly manner, of trying to avoid serving in the people's national defense movement. Instantaneously, Heisenberg realized with dazzling clarity that this was a matter of life and death. The SS was quick about hanging such "deserters" from the nearest tree. The sentry looked through the papers that Heisenberg had issued to himself, and signaled him to go inside and talk with the commanding officer. That had to be prevented. Heisenberg pulled a pack of American cigarettes he had acquired by chance the previous day out of his pocket and offered them to the SS man with the words: "I'm sure you haven't smoked a good cigarette in quite a while — here, take these!" The man did, indeed, take the cigarettes and let Heisenberg continue on his way. [The consequences, had he not been a smoker, would have been too terrible to imagine.]

The readiness to stake everything on one card in decisive moments was a feature of Heisenberg's that allowed him to penetrate into completely new, never before contemplated areas of science. He himself once expressed it like this: "A scientist keeps finding himself in Columbus' position, who had the courage to leave all habitable land behind himself in the almost insane hope of finding land anew on the other side of the oceans."

But now it was not a matter of survival, such as in the

incident with the SS sentry; for Heisenberg it was rather a question of the "truth of science." He wanted to assure that this truth could prevail against lies and libel, so that he could continue teaching it to his young students, for whom he felt responsible. For this he needed to be rehabilitated. He hoped to be able to achieve this by proving that scientific truth stands above race and ideology, that the theory of relativity is accurate, regardless of the fact whether Einstein was a "Jew" or not. In addition, he hoped to be able to demonstrate the despicable and slanderous machinations of his enemies, Lenard and Stark, and to put an end to their derogatory influence. And so, he dared to turn directly to the center of power.

To assure that his letter was really received by Himmler and not lost somewhere in the lower echelons, as is so often the fate of such letters, Heisenberg made use of a method that was probably as problematic as it was effective. He asked his mother to give the letter to the old Mrs. Himmler so that she could pass it on to her son. To make this understandable, I once again need to leaf through some family history.

Heisenberg's maternal grandfather was Rector Wecklein of the Max Gymnasium in Munich, a feared, strict teacher. Rector Wecklein had a circle of colleagues who met from time to time, took walks in the Isartal, and discussed pertinent questions of matters relating to school. Headmaster Himmler was also a member of this circle. Thus, there was a loose connection to the Himmlers. They had three sons. The first was a capable lad; he was a good musician engaged in a respectable position, as was the third son. The middle son, on the other hand, had professional problems. Following the *Abitur*, he had studied agriculture. The poultry farm his parents finally bought him did not prosper. This middle son became the "German head of the SS of the *Reich*."

My mother-in-law was a religious woman and a bitter enemy of the Nazis. But anxious about the fate of her son

and worried for our young, newly formed family, she called
up Mrs. Himmler and asked if she might visit her. Mrs.
Himmler still lived in an old, properly middle class apartment building, a crucifix in a corner of the living room,
with fresh flowers arranged in front of it. The old woman
listened to everything, took the letter for her son, and kept
shaking her head somewhat sadly and with trepidation at
what my mother-in-law was relating to her. Then she said:
"If my little Heinrich knew of this, he would certainly put
an end to these slanderous accusations. I will talk to him
and give him the letter from our son," she resolutely added.
Just as my mother-in-law was about to leave, Mrs. Himmler
turned to her again and asked, with fear in her eyes: "Or
do you think, Mrs. Heisenberg, that my little Heinrich might
not be on the right path after all?" My mother-in-law shud-
dered with fearful horror as she related this macabre scene
to us.

In his letter to Himmler Heisenberg speaks of the "restoration of his honor"; that strikes an odd chord with our
present-day, more sensible ideas. I will try to explain what
he meant with this phrase. Heisenberg was not talking about
the traditional academic concept of honor that came into
being in the wake of an exaggerated personality cult, and
has become the object of so many caricatures — he had
outgrown this concept that, in Germany, was especially
strong in the student fraternities with which he had never
had anything to do, and he only made jokes about. He
would frequently, and with great amusement, talk about
the duel, to be fought with pistols, he had been challenged
to by an elder colleague when he had come to Leipzig as
a very young professor. He had not greeted the man on
the street! Much amused, he wrote to the angered professor,
saying how sorry he was not to have recognized him. This
happened to him frequently, and had not been intended
as an insult. But if he insisted, he, Heisenberg, would accept the challenge, only that he would need a good six

weeks' time to learn how to shoot a pistol. As far as I can remember, he himself always remained calm in the face of the insults that were often being directed at him from one direction or the other and, on occasion, would quote, in a slightly modified form, from Hauptmann's 'Florian Geyer': *Mir ist schon mehr Pech und Schwefel über meine Rüstung gelaufen."**

But in this case, Heisenberg very tangibly meant the recognition of his position as a renowned scientist with "honor." "I have no desire to live here as a second-class person," he wrote in a depressed letter to Sommerfeld on April 14, 1938, in which it becomes clear once again how seriously he was thinking about the question of emigration. No, he had to achieve recognition of his quality as a physicist, as an expert in his scientific field, as an authority recognized in the world as well as in Germany. The expression "honor" was of great importance to the Nazis; it was the language they understood. He was calculating on this, and he used it like a chess piece, positioning it so as to obtain a victory over the philistine spirit of "German physics," a victory that Lenard and Stark, along with all their allies, would have to abide by.

Heisenberg was frequently summoned to "hearings" in Berlin, where he was also questioned in the notorious prison in the Prinz-Albert-Strasse. He never talked about it very much, but I remember that he returned home from these hearings exhausted and distressed. Not that anything terrible actually happened to him. There was an "expert" present at these sessions, a man from the SS, who was a physicist and had attended some of Heisenberg's lectures in the earlier days. This man was well disposed toward him and saw to it that the hearings were conducted relatively factually, and that Heisenberg was spared from vulgar treat-

---

*The English saying, loosely adapted: "I have been through more hell and high water than that" would be an approximation of the German expression.

ment. It became clear to him that it could have been different, when he saw a poster on the wall of the interrogation room with the logo: "Breathe deeply and calmly." This jolted his fantasy, and he saw the tortured faces of the others being interrogated quite differently than he, and for whom there would be no return to normal life.

In spite of his strong and healthy appearance, Heisenberg was extremely sensitive and delicate. "He looked like a simple farm boy, with short, light hair, clear, bright eyes, and a radiant expression on his face," Max Born later reported on his first impressions. And he continues: "His unbelievable quickness and precision of understanding enabled him to regularly accomplish an incredible amount of work with very little effort." But this intense engagement in his work, the dynamics of his genius, also took their toll. His bad allergies, plaguing him for long stretches of the summer, and the bodily discomforts connected with them, were the physical expression of this great sensitivity. Emotionally, this manifested itself in his need for harmony, a harmony assuring him of the inner freedom necessary for his work. The opposite pole, however, was fear, fear of the chaos that could engulf him. No one suspected this behind his open, clear, and decisive bearing. But in truth, it was a deeply rooted fear of losing his autonomy, of becoming subjected to others, a fear, as well, of being tormented, and a fear of great pain. That was the price he had to pay for his sensitivity and his talent. This has to be understood, to appreciate why Heisenberg had no doubt that it was his right to protect himself from the criminal violence of his enemies through certain inconsequential compromises that harmed no one. An idle lifting of a hand meant nothing to him; he thought it ridiculous to get upset about a *Heil Hitler* under an official or a semi-official letter. He made compromises when it was a matter of assuring something he thought to be important, especially if it seemed necessary to protect himself from blatant threats to his existence. He could only maintain his inner composure, menaced by the deep horror called forth in him through the crimes of the Nazis and the war, by contrasting it with the clear world of science in which he saw the eternal verities that would outlast the terrors of those times. In this area, compromise was not possible. His bravery was of an intellectual nature, and to him the compromises were like the ripples on the surface of an ocean, though it could easily swallow you.

The investigation that Heisenberg had brought on himself dragged on for months, despite what seemed to be a relatively favorable constellation for him. There can be no doubt that Himmler's mother spoke with her son and gave him the letter. This also explains why the investigation was finally terminated somewhat abruptly. A meeting between Himmler and Heisenberg never took place, even though Himmler repeatedly made plans for one, and there are documents mentioning them. The worldwide fame of the young Heisenberg may also have influenced Himmler's decision. *One* Ossietzky may have been enough even for Himmler! In addition, Himmler entertained hopes, as can be seen in his letters, that he could win Heisenberg over for his abstruse plans concerning glacial cosmogony, etc. In any case, on July 21, 1938, and thus one year after the publication of the article in the *Schwarze Korps*, Heisenberg received a letter from Himmler with the following message: "Precisely because you were recommended to me by my family, I caused your case to be examined with special care and intensity. I take pleasure in being able to inform you that I do not approve of the attack made against you in the article in the *Schwarze Korps*, and that I have ensured that there will be no further attacks on your person..."

On the same day he wrote to Heydrich: "Please...clarify this whole matter with the league of students as well as with the student leadership of the *Reich*, since...we cannot afford to lose or silence this man, who is still relatively young and who can educate the coming generation..." This letter makes the extent of Heisenberg's peril drastically clear.

And so, partial success had been achieved. Heisenberg had now been given legitimacy to resume the teaching of modern physics as he saw fit. This was of inestimable importance for all the other physicists teaching at German universities, as well as for the whole generation of young physicists that was coming of age, and for which Heisenberg felt himself responsible.

Even so, Heisenberg was not able to reclaim the teaching position in Munich either for himself or for some other respected physicist. The position of Sommerfeld's successor was filled by a completely unknown physicist by the name of Wilhelm Müller, of whom Prandtl wrote in Göttingen: "Mr. Müller contributes nothing, but absolutely nothing to theoretical physics. Instead, he has published a most polemic study program that can only be designated as a sabotage of a subject that is indispensable for the further development of technology." [Unpublished memorandum of May 1941.]

The descent into "the cave of the monster," as we sometimes called it, had cost Heisenberg much of his strength. Had it been worth it? I do not want to judge in these memoirs, only report and correct misconceptions. "German Physics" had not been condemned to death — that happened later on in a completely different context.

chapter four

EMIGRATION OFFERS
FROM ABROAD

As we have seen, Heisenberg's life during the years before
the war was characterized by the tiring battle against the
dictatorship of the philistine spirit prevailing in the sciences,
and the effort to continue his life and teaching at the institute in Leipzig with as little disruption as possible. It had
become difficult to maintain intensive scientific activity after
the heavy blood-letting caused by the expulsion of so many
highly talented Jewish students and colleagues and a great
number of his foreign pupils. He repeatedly tried to sign
up competent, respected colleagues for his institute. As late
as 1934, he attempted to convince the Dutch physicist, Professor Casimir, whom he knew and respected from his days
in Copenhagen, to occupy the chair for experimental physics
in Leipzig. Predictably, and to his deep regret, his offer was
rejected. Others reacted the same way as Casimir.
Heisenberg suffered from this gloomy condition of ever
increasing isolation, struggling against it in vain. Following a visit to Copenhagen, he wrote to Niels Bohr on July
25, 1936: "...I still think back to the enjoyable time in
Copenhagen with great pleasure, and I heartily thank you
and the others for your hospitality. During the last few
years, the occupation of a physicist has become a very lonely business here, and so it is always a cause for celebration
to be able to share in the full life of science for a few days
with your group."
   Nevertheless, Heisenberg once again succeeded in gathering around him a number of talented and dependable young
people. Among the many names of that period I can
remember Flügge, Volz, Dolch, the Yugoslav Supec, the

two Japanese Tomonaga and Watanabe, and, above all, that inseparable trio: Harald Wergeland from Norway, Berndt Olaf Grönblom from Finland, and Hans Euler, a young German communist, who was engaged to a girl who had fled to Switzerland because of her Jewish parentage. After the departure of his assistant Felix Bloch, which Heisenberg had to accept with a heavy heart, he had given the post to Hans Euler, as he was a highly talented physicist with a discriminating mind, but also, to be able to protect him better, as he had a great respect for him as a human being, and to offer him a means of subsistence; for Euler had no resources and was very much endangered by his quite well known political attitudes. In chapter fourteen of his autobiography, Heisenberg describes the fate of Hans Euler, with whom we both developed a strong bond of friendship during these trying years. It was always extremely painful to him that he had not been able to keep him alive through the war.

There were still some friends at the university he could rely on, mainly Bonhoeffer, Hund, van der Waerden, as well as others, although it was becoming increasingly clear that everyone now had to look out for himself and that the political situation was sowing distrust between people. It was the time of the political joke; it could be used to sound out the opinion of others while concealing your own, creating a certain amount of mutual solidarity among those suffering from the times. The increasing difficulties of life created a dependency on mutual help, which in turn also brought out a certain solidarity. Those profiting from the times and demonstratively showing off their membership in the group of the parvenues were excluded; they were avoided and shunned whenever possible.

In these years, Heisenberg was offered several positions at foreign universities. One offer came from Australia; that was no temptation. The only one he seriously considered was the call he received from Columbia University in 1937.

He had real friends there, and that attracted him. On August 31, 1938, he wrote to Sommerfeld: "As early as last summer, Professor Betz of Columbia University presented me with an invitation to go there, either for one semester or permanently." He continues: "In response to this letter I wrote that I preferred to stay in Germany, and that for the time being, I would only come for a short period. For this visit, I have been thinking about the period from February to May, but as yet have not even confirmed that..." A letter to Bohr makes it clear that this decision subjected him to torturous uncertainty. He wrote he almost wished that he had no other choice than to emigrate, that, as with so many of his friends, fate would decide this question for him. But the opposite was the case: his marriage made the answer especially difficult. For Heisenberg, marriage meant more than being a couple, it meant establishing a family. And he still hoped that, finally, his children would be able to grow up in Germany, in the Germany he had loved and in whose hidden existence he still believed deep down in his heart. At the time he had not given up all hope that something could yet change from within. Not that the course charted by the Nazis could possibly be changed — it bore Hitler's mark too clearly — but the army held the factual means of power, and a successful resistance from within its ranks was at least imaginable, and a final hope. Did we not know enough names of officers who were equally shocked by events in Germany as ourselves, and whose courage and decency were beyond question? How often we discussed such a possibility, a putsch of the military, the army that would sweep away the Nazi government, that would not be abused in a war in Hitler's name! This hope had not been buried yet. Moreover, Heisenberg was personally much too involved in his fight against "German physics"; his departure would have meant a fateful victory for this aberrant teaching, and would have endangered the existence of his students and colleagues in theoretical physics. This is probably what ultimately led to his declining the offer.

At this point in time, his reasons for deciding not to emigrate to the United States were certainly more emotionally colored than they were fully thought through. The situation was different when Columbia University repeated its offer again in 1938/39, and he travelled to New York and Chicago for two months in mid May of 1939. By then, all hope of a favorable change for the better in European politics had vanished. The German troops invading Czechoslovakia made it clear that Hitler was willing to risk a war. During his brief visit to England, Heisenberg had seen that "rearmament was being undertaken with the utmost energy, work going on by day and by night," so that no doubt remained that, in the future, Europe would take Hitler up on his challenge. The persecution of the Jews, as well, had reached dimensions that deeply horrified us all; the callous and brutal events of November 9, 1938, during the *Reichskristallnacht*,* paralyzed us with fear. The terror created an oppressive and despondent atmosphere. There were no more doubts: war was inevitable.

Before Heisenberg left for the U.S.A., we decided to travel to Badenweiler for two weeks. I was expecting my third child and was desperately in need of rest. In Badenweiler, on clear days, we could see the graciously flowing lines of the Vosges Mountains across the Rhine, a wonderful view, but it saddened my heart. Would war once again ravage this beautiful countryside? The sight deeply depressed Heisenberg. He commented on it later: "At the sight of the Vosges Mountains, I was gripped by a vision of war and destruction, of the inevitable war we were moving toward." He was so submerged in his problems that he would not speak to me for days on end, and I became wholly despairing as to what to do. I was unable to guide

Reichskristallnacht: a pogrom organized by the National Socialists against German Jews.

him out of his inner darkness, nor did I understand the reasons for this total reserve. Now I do. He was grappling with the question of what to do, how to decide. Did he owe it to his young family to emigrate, to take them somewhere safe? He felt he actually wanted to stay in Germany, but was that responsible toward the family that, after all, he himself had wanted? But if he left, what would become of his students, his friends, of all the others he felt close to? What of his responsibility toward them? Would we even be able to survive in this madhouse? And was there any sense in staying? In this conflict, he decided against the family, against its safety, against the simpler and more pleasant life; this was the reason why he could not talk to me. In a certain sense, he decided against me at the time. The fact that I never reproached him for this moved him to write that I was "brave" in his book, which never seemed to me to be one of my characteristics, and it always puzzled me, until I had understood his reasoning.

Heisenberg was completely aware that with this decision he was also going to tarnish his reputation. He knew that he would not be able to avoid compromises. He writes about it in his book: "At the beginning of a lecture you had to raise your hand, so as to oblige the forms demanded by the National Socialist Party...Official letters had to be signed with a *Heil Hitler.* This was already much less pleasant, but fortunately, there were not many letters of this type to write, and in any case this salutation had the undertone: 'I don't want to have anything to do with you.' You had to participate in celebrations and rallies. But it would not doubt be possible to avoid these responsibilities. Every single step, seen individually, could perhaps still be explained. But there would be many steps to take." Yes, there would be guilt to shoulder, but was there not also guilt involved in leaving?

At the time we certainly did not have a clear idea of what was to happen. Here, too, we were not capable of estimat-

; the extent of the horror. We clung to the hope that the population would not go along with another war, that the army would dispose of the hated regime in reaction to serious, and unavoidable, setbacks, that, in case England entered into the war, Hitler would perhaps retreat from the final consequence, and all the other vague wishful thinking floating around. But these were probably more my thoughts; Heisenberg had already taken a deep look into the abyss in Badenweiler. That is where he had decided to stay in Germany. Why had he made that decision, what had been his reasons?

Heisenberg travelled to America again in the early summer of 1939. The reason for this trip was to ensure himself of his friends there, and to reiterate the motivation of his decision to them. At the time, he still firmly believed in the "world community of physicists" he spoke about so often, and he was totally unaware that he was excluding himself from this community through his decision to stay in Germany. Never having identified himself with the policies of the Nazis, he was of the firm conviction that old friendships could outlast political differences. Fermi, with whom he discussed his intention of staying in Germany, demonstrated a degree of undersanding for his decision, although, ultimately, he did not agree; in New York, Pegram was not convinced by his arguments, and most of his American friends reacted the same way; it was a painful experience for Heisenberg.

It is a frequent assumption that Heisenberg's decision to stay in Germany was influenced by his belief in, and hope for, a German victory. Those who mistrusted him thought that he must have felt himself obliged to fight for this victory with all his ability; — after all, it was known that Heisenberg was a patriot who loved his country — and what would be more natural than to assume he hoped for his country's triumph? But in reality, Heisenberg never for one moment believed that Germany could win its malign and criminal war; it had to lose. Even when appearances could have pointed in another direction, when the German troops were covering themselves with laurel following their occupation of Paris, and the whole world was trembling at the thought that Germany actually could win this war, even then he was not diverted. I remember having asked him at the time whether there might not be a possibility that Hitler could win the war after all. Heisenberg was very calm and assured, and he said: "He will not win. Once an English diplomat was asked how the war would end. He put it this way: The German armies will cut through Europe like a knife through butter; but at some point, they will lose a battle, and their armies will come to a stop. When that happens, they have lost the war!' That is how it will be." And then he added: "Be assured, that is what will happen!" We did not dare have this kind of conversation at home at the time. We were sure that we were being monitored, and we really felt safe only on the street. While talking about these matters, we were walking along Nauenhofer Strasse, under blossoming Japanese cherry trees, on a fragrant spring night.

However, the question whether Heisenberg always thought like this still needs to be asked. Was it his opinion back then in the spring of 1939 when he decided to stay in Germany? Or did he really believe in a victory of Germany, this belief influencing his decision? There are statements of his, made in the United States in the summer of 1939, demonstrating that he thought Germany to be perhaps the strongest military nation in Europe, and thus he did not think a German victory within the European framework to be impossible. I know he thought this. But Heisenberg never doubted in the least that the United States would not tolerate a German victory. And Germany stood no chance against America. It was even more obvious when considering the immeasurably large Russia on the other border of Germany; it had already bled the French army to death! A single glance at a world globe showed the real distribution of power, and Germany's military capability obviously stood no chance. But dare he express thoughts like these? He camouflaged what he wanted to say, expressing himself indirectly, but his friends in the United States did not understand. His remarks about the strength of Germany alarmed them. Only much later did some of them understand what he had meant to say: "So then it is up to you what will become of Europe," and that he had been passing on a share of the responsibility. And the excitement with which Heisenberg awaited the Americans' entry into the war, and how deeply he was disturbed by the fact that they hesitated for so long!.

This should make it apparent that the two statements contain no contradiction. They are differing viewpoints of the same situation, specifically, that Germany did not have the slightest chance of winning the war. And that, on the other hand, means that it was also clear to Heisenberg that he would be on the side of the loser if he stayed in Germany.

In 1939, the emotional-romantic ethos that, perhaps, had influenced his decision less than two years ago had now given way to a much larger comprehension. At most, it still subconsciously influenced him. Now he had reached simple decisions, dictated by his moral code: his place really was in Germany and he could not just get up and leave, if Germany became enmeshed in guilt, misfortune, and even crime. It was the conviction that it was not possible simply to look out for one's own safety and abandon the less fortunate to their fate. He felt himself accountable to all who were his friends, his allies in outlook, and even to the many unknown people as desperate, oppressed, and without hope as he, who could not escape destiny. He had been deeply torn by his duty to secure the safety of his family. But he had decided against it, and had resolved to subject me to the coming catastrophe. And he bought us the cottage on the Walchensee, hoping that the family would be able to

hide out there and survive the war. In fact, I did liv
with the children for almost four years.

Heisenberg never believed that his leaving Germany
would have the slightest effect; nothing would have changed
if he had left his university position and Germany, nothing
other than that he would have saved his reputation. But
in exchange, he would have abandoned his friends and his
students, his family in general, physics, and with it, his colleagues, just to save himself. It was a thought he could not
bear, and it would have doubtlessly made him despondent,
as well as placing a heavy burden on his conscience.

But there were two other serious reasons that had moved
him to make his decision as he had, and not otherwise.
The first was that he suspected that he would, if he were
to emigrate, be obligated to return the proferred hospitality by investing all of his energies in the development of,
probably, nuclear weapons. Since the discovery of nuclear
fission by Otto Hahn, he had been aware that this was at
least in the realm of possibility. But the notion of having
to collaborate in the development of an atomic bomb, possibly then to be used on Germany, causing unimaginable
death and destruction, was a nightmare to him; he would
not do it. And the thought that his presence in America
could have the least influence on the defeat of Nazi Germany struck him as ludicrous. After all, he was absolutely
certain that he was in no way indispensable to the Americans. He knew all too well how many brilliant physicists
were working in the United States: Fermi, Teller, Bloch,
Bethe, Weisskopf; most of them had been colleagues or students of his. In Germany, he hoped, he would have the
liberty of deciding what he would do; there he would only
be responsible to his own conscience. That he could also
be put under extreme pressure in Germany, that he could
even lose his life, was part of his calculation. But basically, Heisenberg was an optimist, and he was not willing
to submerge himself in dismal speculations.

It is impossible to say just how clearly he had developed all these thoughts by the spring of 1939, when he made his decision to stay in Germany; I myself cannot reconstruct when he spoke to me about it in this way for the first time. Of course, he had immediately recognized the possibility of a terrible weapon in the fission of uranium. But just to what extent, how quickly, and with what expenditures the possibility could be realized was probably not even clear to the American physicists at the time.

There is also no doubt that Heisenberg thought the war would be over more quickly than was actually the case. He could not imagine that a people, inwardly thus stifled and oppressed, would have the strength to endure for so long. He had reckoned with two years, thinking that the reserves of the population would then be exhausted. He had probably underestimated the aftereffects of the first great successes; and likewise, he had probably also underestimated the truth of the acknowledged concept that a war against an exterior enemy is capable of mobilizing inner resources to a startling degree. Who actually knew how much the poulation had been weakened by the terrorist regime of National Socialism? With his friend Bonhoeffer, Heisenberg would sometimes argue about the percentage of those swimming on the wave of success of the National Socialists, who had already assimilated the irrational poison of an overblown selfconsciousness, versus those who suffered under its terrorism and were morally weakened, even broken, or actually living in inner resistance to it. The former held the opinion that up to 80 percent of the population backed Hitler and his government, identified with it, and were willing to sacrifice everything for it. By contrast, Heisenberg thought that it was at most 20 percent. Given these uncertainty factors, it was almost impossible for anyone in Germany to judge how long this small country would be capable of holding its own in the challenge it had thrust at such a large part of the world.

If Heisenberg repeatedly said that he wanted to stay in Germany because he was needed there, he did not mean that he wanted to build atom bombs for Germany — something, in fact, that he did not do. Rather, he meant the commitment he had made to staying: to save "islands of stability" for the time of a new beginning. In a letter he wrote to me on January 25, 1946, shortly after having returned to Germany from his internment in England, he says: "Since 1933 it had been clear to me that a terrible tragedy was unfolding; only I could not imagine its extent or end. At the time, I had stayed here to be here when it was over, and to help. That is precisely what I told my friends during the summer of '39, and the best among them understood it well...In any case, during the coming years I want to aid the rebuilding here...since it has to be possible to reawaken something of the lively intellectual life of the 20's..." That was what gave him the purpose and the strength to survive.

At the time of the outbreak of the war, Heisenberg most certainly had strong feelings of guilt toward his country. He felt that he had not been politically aware enough, that he had been dreaming in that he had thought only about physics, or had fled into the romanticism of a carefree life in nature, instead of alertly putting all his energies to use to mobilize moral resistance to Hitler. And now it was too late, and there was nothing left to do but to save what could still be saved. The only possibility now was to wait until the "storm had passed," as he would occasionally put it. And if you survived the storm, it was time to start from the beginning, to try to reconstruct a new order with a new and better spirit. He could not do this alone. So he had to attempt to save people with a similar outlook for the new beginning. That was the other reason for his decision. He was completely absorbed by these ideas. During the war, he was already thinking about what had to be done, later on, when everything would be over with, so as not to repeat the old mistakes. In chapter 15 of his book, *Physics and Beyond*, Heisenberg very vividly describes how even when, together with Butenandt, he barely survived the first heavy bombing of Berlin while in the Ministry of Aviation, his thoughts had been concentrated on the future. They had to climb out over mountains of rubble and then they hurried through the burning city, towards Dahlem, driven by anxiety about conditions at home. Heisenberg, who at the time was living with his parents-in-law in Steglitz on the Fichteberg, was being visited by his two eldest children; and, while he and Butenandt wiped the burning phosphorus from their shoes time and again, they saw the red glow in the sky over the burning Steglitz, and they talked about the time after the war, what would have to be done then, and what this new order they would help to build and for which they would be responsible would look like.

Reflecting on these matters, Heisenberg did not lose himself in general political speculations. When he involved himself personally with the problems of reconstruction, he was thinking about the revival of scientific life, and his thoughts centered on the responsibilities of a scientist, on communications between peoples through science and technology, and he pondered the possibility of creating an advisory board of critical and reasoned scientists as an aid for the political process, and as a controlling organ. In those days the idea of a research council began to take shape in his mind; it would be a panel of scientists, attached to the government. The importance of his commitment to building a new and better order for the future, an order that would not be as easily perverted, is hardly possible to overstate for his decision to stay in Germany.

chapter five

WAR AND THE JOURNEY
TO COPENHAGEN

In August of 1939, Heisenberg returned to Europe aboard
the last ship from America. It was almost empty. After all,
who returned to the madhouse in Germany of his own ac-
cord? It is easy to understand why the Americans formed
the wrong opinion on this move, and decided Heisenberg
was a hidden Nazi, after all.

Meanwhile, the children and I had moved into the cottage we had purchased from Corinth's daughter during the
spring in Urfeld on the Walchensee. At the time, the latter
was living in Hamburg with her husband, and was thinking about leaving Germany for good. On account of the
children, we planned to spend large parts of the summer
in Urfeld; during the winter we all intended to live together
in Leipzig. At the time, Urfeld was a quiet resort, consisting
of about ten houses and two hotels, all inhabited by non-
locals. Our cottage was situated high up on a slope, without
a driveway and completely hidden by the trees. There we
received word about the outbreak of the war, news that
deep down I had anxiously been expecting all summer.
Nevertheless, reality hit me like a bolt of lightning, and it
took me days to regain my composure. As Heisenberg
already had an assignment order for the first day of mobil-
ization in his pocket, we expected him to be drafted any
time. Under these circumstances, we sought to order our
affairs as much as possible in the event that he might not
return.

Heisenberg had an assignment order, since he had participated in two short military training programs generally
required at the time for men over 35 years of age. In 1936

had been in Memmingen for two months, and in 1938, in Sonthofen with the mountain pioneers. He could have probably obtained a release from this duty if he had taken the trouble. The short training programs of the reserves were subjected to the kind of forced volunteerism that was so typical of the Nazi period and that the regime used so effectively to exert pressure on the individual. Furthermore, Heisenberg thought he could demonstrate his loyalty to the state without too many compromises by fulfilling his duty in this manner. He also hoped that he could obtain a certain amount of protection from the attacks directed at him by the ranks of the Party. Besides, he was not really averse to such an exercise. He saw it as physical fitness training, and the rough life in the mountains together with simple people, with whom he made contact easily and where he found acknowledgement, even gave him some pleasure. Thus the few weeks of living as a soldier in the mountains were no great sacrifice to him. But during his service in Sonthofen, he had been involved in the "Czech crisis," and the train that was to transport his unit to action on the Czech border should things get serious, had been kept in readiness for days. The worst had mercifully passed us by then; but now things were irrevocable and, it appeared, there was no way out. All that remained to do was to survive the coming catastrophe.

Instead of being called to the front, as we had expected, Heisenberg was informed a few days later that he should report to the Army Ordnance Department without delay. There he was given the order, as he described it later in an interview with David Irving,* "to work on the question of the utilization of atomic energy together with a group of other physicists." He was told: "Give the matter some thought, whether, under the presently known circumstances of the known characteristics of the fission of uranium, you

---

*Irving is the author of the book, *Der Traum von der Deutschen Atombombe*, Gütersloh, 1976.

think a chain reaction is a real possibility, and if so, then please write down what you think about it." This was the outcome of a meeting in the Army Ordnance Department on September 20, 1939. For us it meant that Heisenberg could stay in Leipzig for the time being, and thus our lives temporarily did not change much. Within two months Heisenberg had already discovered, and once again I quote from the interview with Irving, that "a chain reaction could probably be realized by slowing down the action of the neutrons without simultaneously absorbing them, and that there were actually only two substances that do this very well: heavy water and carbon." Together with his colleague, Professor Robert Döpel, he developed an early form of an atomic reactor during the years 1940-41. All this took place within the very limited space of the rooms of the Leipzig institute: compared with the enormous efforts of the Americans, it hardly appears to be urgently targeted war research, even if this undertaking did prove to be of fundamental importance for the whole later development of reactors in Germany. Once the small test reactor caught fire. This considerably unnerved the responsible physicists, as it was not at all clear that this was not the possible beginning of an uncontrollable chain reaction. But in the end, everything worked out well. Only the firemen who had been called in shook their heads in disbelief at the strangest thing they had ever experienced. And the next day the whole city was whispering about the secret *Wunderwaffe* being built in the Institute for Physics.

The theoretical calculations were making it more and more obvious that an incredibly powerful explosive agent, outdistancing anything hitherto known, could be obtained through atomic chain reactions. But from these calculations, Heisenberg also concluded that the expenditures necessary for such a development would be enormous, and that the path to success could be strewn with countless difficulties that could possibly take years to overcome. This was the

state of affairs when Speer, at the time the Minister responsible for arms development, called together a secret meeting
in the Harneck house in Berlin-Dahlem on June 4, 1942,
where Heisenberg was to report on his research. He con-
scientiously complied. According to his own statement, he
discussed during the meeting the possibility of building
atomic reactors that would be capable of producing tremendous amounts of energy. Later, in his interview with
David Irving, he said that he did not mention that a
byproduct of the generation of energy was plutonium, with
which an atomic bomb could be made. He added: "We
wanted to keep things as small as possible!" When asked
by Commanding General Milch about the possibility of
building an atom bomb — at the time it was called a
"uranium bomb" — and how large such a bomb would have
to be to destroy a city like London, he answered: "The size
of a pineapple." Dr. Telschow, the General Secretary of the
then Kaiser Wilhelm Society, reported that Heisenberg told
him after the meeting: "All the processes we presently know
to construct a uranium bomb are so incredibly expensive
that it would perhaps take many years and require an enormous technical expenditure, costing billions." Thus,
Heisenberg had retreated behind the difficulty of the production of the bomb and its expense. He did nothing to
try and convince the responsible people in the government
to seriously attempt to build the bomb. No doubt, if he
had wanted to achieve this he could have. In his memoirs,
Albert Speer writes about the outcome of this meeting:
"Based on the suggestions of the nuclear physicists, we
decided against the production of an atomic bomb as early
as the fall of 1942, after it had, in response to my repeated
inquiry regarding dates, been explained to me that it could
not be expected before a period of three to four years. By
then the war already had to have been decided. Instead,
I gave permission to develop an energy-producing uranium
reactor for powering machines, which the Command of the

Navy wanted to use in submarines."

It is obvious that Heisenberg had to maintain strict secrecy at the time. Any mention of his work would have been high treason. This is the reason why he hardly spoke to me about this matter during the first years and why I had little knowledge as to how he was dealing with the problem of the bomb. There is no doubt that the purely physical problem challenged him. But he did not talk about the consequences of his actions and the problems of the bomb until later, when we had already moved to Urfeld permanently, and there was hardly any danger of my inadvertently saying the wrong thing. We were much too adjusted to the deadly peril of our situation for this to have been a real temptation to me, anyway. But for him, such a conversation meant relaxation and release.

The meeting on June 4, 1942, during the course of which the fate of the uranium project had substantially been decided, was also the turning point in our dialogue. I clearly remember that he told me details of the meeting. Our discussion about the atomic bomb, the one I still remember most vividly, must have been during this period as well. We were in the living room of our house in Urfeld. We felt safe there, and had the feeling that we could talk freely. Heisenberg had come to visit us for a weekend and was sitting there relaxed and free of tension. He told me that he was now sure that an atomic bomb could be built. I was deeply shocked and asked him: "And what will you do if you are forced to construct atom bombs?" Heisenberg answered very calmly, so that it became clear that he had thought this question through untold times: "Don't worry. We will not build atomic bombs. The development of a bomb is a gigantic project. It would probably take years. And here an ordinance of Hitler's, that no project is to be started that would take longer than half a year, does us a service. Secondly, we have an ally in the economic situation of our country, given the constant bombings and the completely taxed possi-

f our war industry. There is just not enough room!
nomy is no longer capable of the technical expen-
hat would be necessary to build a bomb. All other
war related work and projects would have to be interrupted,
so as to free the energies for this large and difficult project;
the impossibility of that is plain to see. We are lucky that
things are as they are, and that these facts spare us from
having to make a real decision. It is being made for us,
and for this we are grateful. And then," he added, "even
if Hitler were to force us to build the atomic bomb — I
am of the opinion that a person cannot be forced to make
inventions, or new scientific developments, if he doesn't
want to. Creative thought requires a certain free expanse,
or else it can't develop, and all pressure or force will only
hinder, if not stifle, this creativity and will lead it away
from its goal." This is how Heisenberg saw the problem,
and from this conviction he drew the moral strength to continue along his path calmly. Any kind of speculation that
would have happened if . . . is completely idle. No one can
tell, and the facts have proved him right. The Americans
succeeded in completing the atomic bomb only after the
war in Germany had already come to an end; and this with
an incomparable effort that the Germans could never have
matched. Not only did the "Manhattan Project" — the cover
name for the organization working on the development of
the bomb in the U.S.A. — have the still unexhausted economic resources of a vast and wealthy country at its disposal, it also could work with a great number of the most
brilliant thinkers of modern physics and a human asset willing to make the greatest sacrifices in response to the criminal regime of its opponent. Anything Germany would have
been able to achieve would have demanded the greatest
of efforts, as well as an untapped source of energies no
longer available anywhere. Heisenberg had judged the situation completely realistically. He was grateful that that was
the way it was; this situation supported his moral convic-

tion and decision so meaningfully that the issue of the atomic bomb could never even come to a head. All this made it possible for him to work without large economic expenditures on a small experimental reactor during the war. And it also explains why there was never a German attempt to build an atomic bomb.

During the conversation in Urfeld, Heisenberg's thoughts on the problem of the atomic bomb were already clearly ordered and defined. Previously, however, there had certainly been long stretches, when he and his closest friends had felt trapped and intimidated by the dark ambiguities of their situation, and when their thoughts had not reached the point of clarity reflected in our conversation. They must have lived with them in constant controversy, in anguish, fear, doubt, and agony. Heisenberg's desire to travel to Copenhagen again, to see and talk to Niels Bohr, grew out of this situation. Denmark had been occupied by the Germans in May 1940. Heisenberg had assured that Bohr and his Institute would remain untouched through negotiations with the German authorities in Copenhagen. Thus, a trip to Copenhagen was completely feasible. I remember clearly with what joy and eagerness Heisenberg worked at realizing this desire. No trip to an office, no petition or request was too tiresome for him to reach his objective. He was lonely in Germany. Niels Bohr had become a father figure to him. In Tisvilde, the beautiful vacation home of the Bohrs, he had played with their children and had taken them for rides on a pony wagon; he had gone for long sailing trips on the ocean with Bohr, and Niels had visited him in his ski cottage; together, they had grappled with the problems of physics, and he thought that he could talk about anything with Bohr. Heisenberg's wish to travel to Copenhagen must be understood against the backdrop of his still unbroken trust in his friend and teacher, as well as in the whole "international family of physicists" to whom he felt so deeply connected.

Preparations were finally completed in September of 1941. In spite of their difficult political position, the Bohrs received Heisenberg with great warmth and hospitality — a moving sign of close, personal friendship. The complete failure of their mutual discussion was thus all the more painful.

There has been repeated conjecture about the trip to Copenhagen, and Niels Bohr himself apparently formed an impression of its motivations that was in tragic contradiction to Heisenberg's true intentions. It is not my place to get involved in the difficult and emotionally fraught controversy of the meeting of the two men in Copenhagen. My contribution can only be to report on what I experienced myself, and on how it was presented and reflected in our conversations. Let me state this explicitly.

As with all complicated actions, there are certainly different motivations involved in Heisenberg's decision to travel to Copenhagen. It can be assumed that he had developed a simple wish for conversation and understanding based on his attachment to Niels Bohr. Heisenberg's feelings had always been quite unaffected; and the advice of an older friend, more experienced in human and political affairs, had always been important to him. In connection with this it has been said, however, that Heisenberg wanted Bohr's absolution, so he could work on the atom bomb. Not a word of this is true. Had that really been what he was hoping for, the conversation would have been completely satisfactory, since Bohr did, in a certain sense, actually give him a kind of absolution. He told Heisenberg that he understood completely that one had to use all of one's abilities and energies for one's country in time of war; that was clear and to some extent true. But he failed to see that Heisenberg was talking about completely different matters. In truth, Heisenberg was deeply shocked by Bohr's reply; he had not expected it, and it showed him that they were missing each other's point, no longer understanding each other, and that the discussion was perhaps doomed to failure.

So what was Heisenberg's ultimate concern during these discussions with Bohr? The truth was that Heisenberg saw himself confronted with the spectre of the atomic bomb, and he wanted to signal to Bohr that Germany neither would nor could build a bomb. That was his central motive. He hoped that the Americans, if Bohr could tell them this, would perhaps abandon their own incredibly expensive development. Yes, secretly he even hoped that his message could prevent the use of an atomic bomb on Germany one day. He was constantly tortured by this idea. And he was firmly convinced that the atom bomb would not be decisive for an allied victory; the course of events did, indeed, prove him right. This vague hope was probably the strongest motivation for his trip. But they were not communicating; they did not understand each other. The two men who had been such close friends parted deeply disappointed, and there were no further attempts to contact each other.

Ultimately, the cause of this misunderstanding must be seen in both of them. Bohr was on the side of the country being savagely attacked and occupied. He had to have deep reservations about Heisenberg. Talking to him bordered on collaboration, especially if they had parted in full agreement. This alone was an almost insurmountable barrier. So Bohr could hardly avoid acting on false assumptions. He knew of Heisenberg's love for his country; he knew that golden bridges had been built for Heisenberg's emigration to America, and that he had turned them down. What was more logical than to believe that Heisenberg would do his utmost to assure the victory of his country? Had he himself not told the Americans that he had to stay in Germany, because he was needed there? Now he came as a member of the victorious nation, and Bohr was committed to the side of the allies with all his heart. To him, Nazi Germany was an abomination, in addition to its being a daily threat to him as a "half Jew." He fervently hoped for an allied victory. In this situation, it was probably too difficult for him to understand that Heisenberg's bond to his country and its people was not tantamount to a bond to the regime, and that Heisenberg himself was in the same tragic position as those who abhorred the government as much as he did and were thus incapable of working for the victory of their own nation with all their available energy.

Heisenberg had not been aware of this beforehand. The psychological situation was quite unexpected, and it threw him into a state of confusion and despair. He had quite different problems in talking with Bohr. For him, the greatest problem was that he could not speak freely and openly about his own affairs. He had not taken into consideration that inhabitants of the free world were not used to, much less practiced in, understanding the hidden language used to communicate in a society ruled by terror. For his central discussion with Niels Bohr he had carefully decided to go for a walk, just as we did at home if we wanted to talk about politics. There was no third party, no witness, present. And he still did not feel safe, since every word he said could have been regarded as high treason and cost him his life. A single public quote and his life would have been forfeited. He had just recently experienced how carelessly the spoken and written word was used in the free world, upon receiving a letter from Switzerland saying: "Dear Colleague Heisenberg! I would like to inform you that our colleague G. [name in full] is hiding out in such and such a place under another name. It would be nice if you could do something for him; he is completely destitute." Heisenberg was very upset when he returned home. Naturally the letter had been opened, and not only was he now in danger himself, it was also too late to help G. Heisenberg distrusted the discretion of others, people who had no concept of the methods of total surveillance. He had to prevent the statement, "Heisenberg said that..." So he could not speak directly and openly.

In his book *Physics and Beyond,* Heisenberg tried to reconstruct the conversation. But it was seemingly laced with so many emotions on both sides that his description can only be seen as an abstraction. For all that, it does show how carefully they both formulated and acted, and that Bohr essentially heard only one single sentence: The Germans knew that atomic bombs could be built. He was deeply shaken by this, and his consternation was so great that he lost track of all else. And this put an end to any kind of communication between the two.

To do the situation justice, it also needs to be said that even if Bohr had understood and accepted everything else Heisenberg said, the course of events could still not have been changed. No one in America would have believed Heisenberg; they would probably have viewed it as a clever ruse of the Germans. At that point they were too deeply involved in the gigantic undertaking of developing an atomic bomb, anyway. Once again, events were stronger than the efforts of any one individual. Heisenberg often expressed it with the following image: you can't stop a moving train with your bare hands! Much too much had already been invested in the project; in comparison to those of the German venture, its expenditures were like a giant to a dwarf. Heisenberg had no idea of this; German intelligence did not have the slightest inkling. The tragedy of this episode is that the conversation in Copenhagen probably had precisely the opposite effect of what Heisenberg had hoped for, namely, an intensification of work on the bomb. For Bohr, deeply troubled and in the belief that Heisenberg had perhaps been trying to sound him out, told the Americans that Germany was working on the construction of the bomb, that they knew how to build atomic bombs, and that Heisenberg was the leader of this project. Things had been turned around completely.

Some time after this unfortunate discussion, in January 1944, Heisenberg travelled to Copenhagen again on a quite different mission. In September 1943, Bohr had fled from Denmark, taking a small boat over the sound by night, apparently fearing a threatened pogrom against the Jews. Subsequently, the SS seized his Institute, and his colleague, Dr. Böggild, was arrested on suspicion of "conspiracy with the enemy." Heisenberg had been informed of these events through circuitous routes, and he travelled to Copenhagen immediately and exerted all his influence to secure the freedom of Dr. Böggild as well as of the Institute. After a few days of intense efforts, he managed to change the mind of the commanding officer, Dr. Best. Both Dr. Böggild and the Institute were released from custody, and the Copenhageners could once again resume their work without interference. However, Heisenberg had to pay a price here, too: he had to talk to all of the important people of the occupation forces and the Gestapo, he had to eat and drink with them, and had to move about as though he were one of them; there was no cheaper way.

It took a long time for at least some of the old friendship between Bohr and Heisenberg to rekindle after the war, and for Bohr to lose at least some of the deep mistrust that the war, and especially that unfortunate discussion, had awakened in him. The old, unconditional friendship and the unbroken trust that had existed between them earlier could never be completely restored. This troubled Heisenberg for the rest of his life.

10 Fermi, Debye, Heisenberg, and Bohr during a conference in Rome in front of the institute of physics in the thirties

11 Heisenberg in the thirties

12 Heisenberg in 1931

13 Heisenberg at the Lake Constance shortly
after his wedding, 1937

14 In the mountains with Niels Bohr
(around 1931)

15 In the fall of 1937 in Copenhagen

16 Heisenberg, Max von Laue, and Otto Hahn during the development
    of the Göttingen Institute (1946)

17 The Urfelder house

18 Farm-Hall, the "golden cage" of the ten German nuclear physicists

chapter six

THE KAISER WILHELM INSTITUTE IN BERLIN AND THE LAST YEARS OF THE WAR

Half a year after Heisenberg's failed visit with his friend and teacher Niels Bohr in Copenhagen, he accepted the position as director of the Kaiser Wilhelm Institute in Berlin-Dahlem on April 24, 1942. This was not a surprising step, as it had been initiated well in advance. Heisenberg was severely criticized for having assumed such an important position in the Nazi regime at such a late date. Why did he not silently remain in the background and work out of sight like, for example, his highly esteemed colleague Max von Laue? This position offered only slightly better research possibilities; above all, it meant more prestige, a broader range of influence, and more political responsibility. Besides, what about his other, older colleagues who perhaps had a better claim to the position, such as Professor Bothe from Heidelberg? And finally, was this post not better filled by an experimental physicist rather than by a theoretician? Debye, who had directed the Institute during the last few years before the war, was also a student of Sommerfeld, but had later immersed himself completely in experimental physics, and many felt that someone like Heisenberg, who had distinguished himself through his research in fundamental physics, did not belong in this office. Heisenberg's critics raised all of these questions, and interpreted his decision as further proof that he was acting out of opportunism and ambition, and that he really had wanted to construct an atomic bomb. They argued that he obviously would not have had these chances as a German in America; there he would only have been one among many. And it was indeed easy to understand his actions in this context, for when Heisenberg took over the Institute, the Army Ordnance Department had decided to concentrate all the various groups involved in nuclear research throughout Germany in the Berlin Institute to consolidate its command over the nuclear project. With the Kaiser Wilhelm Institute, Heisenberg was not only taking over the highest scientific research position available in Germany, he simultaneously also became the head of nuclear research there.

What led to this? What was Heisenberg's role in these events? Was it really his ambition that drove him to take over the leadership of the "uranium project," as it was now called? It is necessary to take a closer look at the details to get a clearer and more accurate picture of these events.

But first, a word on the subject of Heisenberg's ambition might be appropriate. There is no doubt that he was ambitious. His teachers in school were already praising his aspirations in his first report cards. And he did come from a family of pedagogues — the talents of his father in this respect had always been acknowledged — and in all probability his grandfather Wecklein, a headmaster himself, who took real pleasure in his talented grandchild, quite consciously encouraged this ambition of young Heisenberg, just as his father did. Sometimes Heisenberg would tell us how the family played guessing games and tested themselves with intellectual problems, and that he had actually always been quicker than the others; or how his father one day gave him a Latin text on number theory, a subject he was interested in at the time, to improve Heisenberg's Latin and not his mathematics, and how brilliantly the young boy had done this. He himself, he told me, was deeply impressed when he discovered a Greek saying above the entrance door of a secondary school, meaning roughly: "Always strive to be the best at whatever you do." And he decided to make this his guiding motto. His great talent made it easy for him to live accordingly; he proceeded almost playfully to be the best in his class in most subjects throughout his schooling. Only in sports did this call for an extra effort. As a boy he tended to be frail and awkward. This annoyed him, and he decided to change it. Every day, as soon as it got dark, he would run a few kilometers around the Luitpold Park, stopwatch in hand, constantly checking his speed. After having kept this up for about three years, he received top grades in sports as well, and henceforth he had no more problems with activities like mountain climbing, hiking, and skiing. Ten years later he was on a lecture tour in Japan and was challenged to a ping-pong game. He lost, not having stood a chance against the practiced Japanese. On the boat trip from Shanghai back to Europe — he had booked it on a Japanese vessel as it was cheaper — he trained systematically and intensively. From then on, amateurs could hardly beat him. He was so gifted that he showed great achievements wherever he set his will — and he derived pleasure from it.

In addition, Heisenberg was industrious and had unbelievable powers of concentration. He always demanded the most of himself, whatever he was engaged in: a game of chess, playing the piano, hiking in the mountains, or skiing; even if he cut some flowers in the garden and arranged them in a large vase, it was done with concentration and care. His bouquets were colorful and exuberant.

I myself did not much care for this ambition in Heisenberg, especially when he applied it to family games. It irritated me to see a certain naive egoism hidden behind his efforts. But it always remained obvious that this display of ambition was something quite different from a demonstration of personal superiority. In his eulogy, Weizsäcker says the following: "He had a competitive, fair, uncontrollable ambition for achievement." Today, it would probably not even be called "ambition"*; rather, he would be described as being highly achievement-motivated. This would also come closer to the truth. Indeed, his ambition was not directed at the acquisition of "honor" at all; he was devoid of vanity. To him, praise, homage, applause, and admiration were relatively unimportant. He was pleased by the recognition of people he valued, whose incorruptible judgement he knew. And he was willing to admire others for their achievements throughout his life. In his later years, his fame almost became a burden that had to be shouldered. Shortly before his death he said to me: "There is a danger that I will be given a state funeral. I don't want that. I want to be carried to my grave surrounded by my family and my closest friends." And that is how it was done.

But in spite of his personal modesty, he was thoroughly aware of his value, and when he thought something to be important, his demands were high. When he arrived in Leipzig as a young professor and moved into the attic apartment of his institute, the first object he put into his otherwise empty residence was a grand piano; it was the best he could find, a Blüthner piano of impeccable quality. It had been specially made for the World Exhibition in Paris, and its tone was more beautiful than that of any other piano he had ever played. This he indulged in. With regard to his personal needs, however, he was remarkably unpretentious. Requesting a house with a garden for his large family when housing was still in short supply after the war certainly troubled his conscience.

In matters of science it was different. Regarding the succession of Sommerfeld, Heisenberg was convinced that the position of his former teacher was his due. He was certain that he would be able to assure the honorable continuation of Sommerfeld's famous school, following his own

---

*German *Ehrgeiz* also means "greed for honor." [Translators' note]

ideas, and he was prepared to use all his energy to make it blossom anew. That was his ambition and his pride. But the situation in Berlin was completely different.

The reasons for Heisenberg's acceptance of the directorship of the Kaiser Wilhelm Institute were the following: as early as 1935 Professor Debye had already left Leipzig to take charge of the Kaiser Wilhelm Institute for Physics, much to Heisenberg's sorrow, since one of his main reasons for going to Leipzig, rather than to Zürich, had been the prospect of working with Debye. On his move to Berlin, Debye had taken along Karl Wirtz, who was a colleague of C. F. Bonhoeffer and a close friend of Heisenberg. At the time Weizsäcker was a member of the Berlin Institute as well. At least according to Heisenberg, Wirtz was in a certain sense the clearest political thinker among his close friends, and an ardent anti-Nazi. After the outbreak of the war, the German government had given Debye, who was Dutch, the choice of either accepting German citizenship and staying on as director, or of leaving Germany. Debye requested leave to give guest lectures in America and did not return to Germany. Thereupon the Army Ordnance Department seized the orphaned Institute and installed Dr. Diebner as acting director. When, in this situation, the Army Ordnance Department decided to consolidate the scattered departments working on nuclear research in Germany and to give them a central command, the Berlin Institute was the obvious choice for its location. Wirtz and his friends recognized the great danger inherent in this. They feared that the Institute, and thus the uranium project, would be permeated with political functionaries; after all, they had seen this happen in other institutes. With this, the leadership would no longer be under the control of a reliable scientist but in the hands of the Party functionaries, and the scientists would lose control of the uranium project. There were also signs that the Berlin Ministry intended to subordinate the Institute to a follower of Lenard and Stark, a certain ministerial director Mentzel.* This ambitious Nazi, driven by his deep rancor, had already attacked Heisenberg and theoretical physics in a most unpleasant manner several times, and he was thought to be unscrupulous and dangerous. In all probability he would have been an even worse misfortune for the Institute, the "uranium project," and the physicists engaged in it than a purely political functionary with only vague ideas about the possibilities of nuclear development. It was quite clear to Wirtz, Weizsäcker, and their friends: such a development would mean the end of any free decisions and any open exchange of ideas in the Institute. The consequences of this type of leadership were unimaginable, especially in such an exposed and central office, and the danger that the decisions relating to the bomb project would slip out of their control and that, counter to all reason, the project would be reinvigorated and forcefully accelerated would become a distinct possibility. Mr. Mentzel would have had no great difficulty in persuading Hitler that he could find his often discussed and quoted *Wunderwaffe* in the "uranium bomb." And this would subject everyone working on the development of the reactor to mortal conflicts and dangers. It had to be prevented.

So, in agreement with his friends, Wirtz decided to hinder such a development through a carefully considered, long-term action. The plan was to increasingly tie Heisenberg to Berlin through a continuous expansion of his responsibilities as an advisor and lecturer at the Institute, the technical, as well as the regular university, and in this way to have him infiltrate the administration. At the Institute there was unanimous agreement that Heisenberg was the right man for the job. Not only was he reliable and fair as a per-

---

*I have not been able to ascertain if the author of the article in the *Völkischer Beobachter*, whose name is given as Menzel, is, as I suspected, identical with the later ministerial director; I have since been assured that he spells himself with a "tz," thus, Mentzel.

son, as the argument went; he had also familiarized himself already with the technical problems of constructing a reactor to such a degree that he could be seen as an expert in the field, even if he was a theoretician, and his main interest had always been to explore the basics and their philosophical consequences. In a report authored by the specialists Wirtz and Häfele in 1961, this is once again clearly stated: ". . . he was seemingly able effortlessly to train himself in a secondary field like reactor technology quickly and comprehensively, and to become the leading thinker for the complete development in this field in Germany for many years."

The whole idea had naturally not arisen over night. And it is not as though one single person masterminded the whole plan. It took shape slowly, and was so convincing that even Dr. Telschow, the administrative director at the Kaiser Wilhelm Society at the time, accepted it. It started with asking Heisenberg to give lectures at the University of Berlin. There was a lack of qualified university teachers everywhere, just as people everywhere were hungry for intellectual sustenance not tarnished by propagandistic lies. So Heisenberg did not hesitate to accept the offer in spite of the twofold burden. Simultaneously, Wirtz involved him in the Kaiser Wilhelm Institute in Berlin as an advisor, where he increasingly became familiar not only with the scientific problems, but was mainly confronted with the many political ramifications of the situation.

I myself was anything but gladdened by this double burden on my husband. Not only did this condition exceed Heisenberg's strength, his body was already beginning to rebel with serious, herpes-like inflammations, but the first devastating large-scale attack on Hamburg had demonstrated what could be expected in the cities. And there was no doubt that Berlin would be one of the main targets of these attacks. But separation and danger were the common lot; after all, it was war. We were no more exempt from it than anyone else.

It was becoming ever more apparent to Heisenberg by now that events were developing such that he would be forced to take over the control of the Berlin Institute. For some time, he balked at the thought and postponed his decision, as Debye was officially still the acting director of the Institute. Even when the administration of the Kaiser Wilhelm Society finally asked him to take over the Berlin Institute as Executive Director, the decision was difficult. He feared the burden of responsibility that he would have to bear if he officially took over the Institute, and, in a very depressed state of mind, he tried to arrive at a clear view of the correct action to take in his conversations with me. Would this high position not put him back in the direct firing line of the Party? And would that not be much more dangerous now than it had been in the past? He was also distressed by the idea that he would possibly have to make even larger concessions to the Nazis and to suffer even greater conflicts of conscience. After all, it was a political position; in other words, in it he would not be able to simply withdraw from the political problems coming his way; he would have to take a stand, and what would be the consequences of that? And then, there was the family to consider. Taking charge of the Institute would mean that he would either be more separated from his family than ever, or he would have to take them along to Berlin — but could he take responsibility for that in view of the bombardment of Berlin increasing from month to month? All these arguments severely depressed him. On the other hand, did a free decision even exist? One thing he recognized clearly: if he wanted to continue along the path he had selected, if he wished to persist in saving young, promising, talented people through the war and the catastrophe and, in addition, to create "islands of stability and freedom," then he could not turn down this position and faint-heartedly give up during this most difficult phase of the war and the Nazi period. It became obvious to him that there was no way out. So he accepted the offer, and was appointed director of the Intitute for Physics on July 1, 1942. The only thing that gave Heisenberg any satisfaction in this matter was the fact that his summons to Berlin had to be viewed as a clear victory of modern physics over "German physics," for if he held the key position at the Kaiser Wilhelm Institute, the highest scientific position for physics in Germany, the phantom of "German physics" was dead. I know how important this was to him.

At the Berlin Institute as well, the goal of the experimental research remained the energy-dispensing reactor, and not the atomic bomb. After all, Speer gave up the whole bomb project in the fall of 1942. And if the government continued to spread rumors about the *Wunderwaffe* with which Germany would finally win the war, they could not possibly have meant the atom bomb: the responsible people knew for a fact that it was not being built. Most probably the talk about the *Wunderwaffe* lacked any basis in reality and was only intended to calm the population and bolster its perseverance. And the small funds available to the reactor project, so often a source of amusement and even ridicule for the Americans, the intentionally modest scale that was so obviously apparent in the cellar of Haigerloch, demonstrate that the scientists in Germany never made any efforts to reach the goal in which "*Wunderwaffe* = atom bomb." To restate it in all clarity, the goal remained the development of the basic principles of reactor construction for the time after the war, when it would be necessary to restore the sciences in Germany and to reestablish connections to international science and technology. And even though the latter goal could only be reached after years of trying, the work on the reactor had not been in vain. Even if it was not possible to have the reactor in Haigerloch operational when the Americans marched in and "conquered" it, later, after the war, when work on the reactor for peace-time use of atomic energy was resumed in Karlsruhe with a greater effort, it became obvious that research could continue on the basis of what had been developed during the war. This was acknowledged by Wirtz and Häfele.

But in Berlin the difficulties for scientific research were increasing daily. There were attacks every night, the sirens shrieked by day as well, and everyone ran for a cellar to seek protection from the bombs. The destruction ate its way further and further into the outer areas of the city. Peaceful Dahlem, surrounded by gardens and the location of the Kaiser Wilhelm Institutes, was not spared, and frequently the whole staff of the Institute, including Heisenberg, had to help with putting out fires in neighboring institutes, cleaning up, performing urgent repairs, or even looking for and rescuing those buried in the rubble after the "all clear" had been sounded.

Up to now, Heisenberg's Institute had been spared heavy damages. Its members unanimously claimed that it was the doing of Saint Florian, of whom a small statue had already been placed in the corridor of the Institute by Debye years before. Saint Florian is the saint who protects people from fire, and now that the bombings were becoming more frequent in Dahlem, his small statue was always decorated with fresh flowers. "Supposedly it will help even if you don't believe in it" — using this famous statement of Niels Bohr, even the greatest skeptics placed their flowers at the feet of the saint. Nevertheless, given the circumstances, progress at the Institute was slow.

In this situation the government issued a decree in the summer of 1943, stating that all important institutions were to be moved out of the city into less endangered areas. This offered the Kaiser Wilhelm Institute the possibility of moving its research out of Berlin. Wirtz and his staff were instructed to find a suitable location for the Institute, and he and Heisenberg quickly agreed on certain conditions that had to be satisfied by the new location. The most important consideration was to be the safety and the probability of the people involved to survive. This entailed, so it was argued, that the new location be situated far enough to the west, so as not to fall into the Russian sphere of influence when the allied troops marched in. On the other hand, it should not be too close to the western industrial centers, as it was assumed that they would be subjected to especially forceful and concentrated attacks. Furthermore, the new location should not only display satisfactory working conditions, it should also offer ample and good living space for the families of the members of the Institute. And above all, it should be situated in a good agricultural region so as to assure food for all these people, for we were convinced that the food supply to the population would collapse soon. These were many and difficult conditions, but the still completely unharmed towns in Swabia with their small industrial plants, partially closed down, proved to be exceptionally well suited for the purpose. After carefully examining the possibilities, the administration of the Institute decided on the small town of Hechingen. The nearby rock caves of Haigerloch, where the reactor was to be placed, offered ideal protection from any air raid. Naturally the relocation of the greater part of the Institute and all the families involved cost, time, and energy, and all hands were needed. In retrospect, it seems almost impossible that it was executed without a serious disturbance. And finally the new abode in the Swabian idyll proved its value, and work on the reactor could be resumed.

In Hechingen, Heisenberg lived in a lovely, large room in the home of a friendly and helpful family. The children and I, however, had no share in the Swabian refuge. We had completely moved to Urfeld at the same time, into the house that we had already bought for this purpose in 1939. Heisenberg was of the opinion that this secluded spot, protected by the mountains, offered his family more security than the proximity of the Institute, where heavy bombardments were, after all, a possibility. Whether or not this was a fortunate decision was always a slight cause for dispute between us. In Urfeld, we were cut off from all the help the people in Hechingen so naturally exchanged. The soil was rocky and barren, and the little that did grow was predictably eaten by the deer. In addition, the farmers harbored an implacable, distrustful stinginess toward us outsiders. In fact, we had serious difficulties, and we waged a grim battle against hunger and sickness. By the time this situation had become clear, it was too late to move to Hechingen, since finding an apartment in Hechingen was an impossibility by then. Heisenberg felt all the more obliged to help us however he could. His letters from this period show how this problem seemed to occupy all his thoughts. As he still had to go to Berlin often, he took advantage of these opportunities to pick large quantities of fruit in the lovely orchard of the Institute that Debye had planted. Then he made preserves with his own hands, or he packed it into large crates to send to our "eagle's nest" in Urfeld, as I liked to call it. However, it usually never arrived, and when it did it was weeks later and either spoiled or broken. He tried to supply us with potatoes and firewood, and made all efforts to obtain the material necessary for the urgent repair of the roof of our cottage. All these were exhausting and involved tasks.

But truthfully, Heisenberg's work was hindered by far greater problems than worrying about his family. The destruction of the heavy water plant in Norway through the daring attack of an English shock troop slowed down work on the reactor considerably, and imposed serious limitations on the delivery of heavy water so necessary for the construction of the reactor. Heisenberg and his team had great difficulties dealing with this new situation. But, in any case, he was no longer striving for a spectacular success. Work in Hechingen was generally carried out in a calm frame of mind. Naturally, he would have been glad to see the reactor operating before the end of the war, and he was always a little hurt by the fact that he had not followed the easier path of using carbon instead of heavy water, because he had relied on the incorrect calculations of another institute. But there were more important things to consider; human problems now took precedence.

Heisenberg had already tried to take care of his endangered colleagues as well as he could in Leipzig, and to find positions for them abroad. At the time he also took in Dr. Gora from Warsaw; he had fled when the Germans marched in and Heisenberg helped him. In Berlin, it was mainly the professors Gans and Rausch von Traubenberg whom he stood up for and saved. He also used his influence to help the Frenchman Cavaillés and the Belgian Cosyns. In the papers he left behind after his death, I found a note on these cases, concluding with the following: "Unfortunately, Cavaillés was shot before my petition with the Foreign Office was successful; or so I was told later on. On the other hand, Cosyns supposedly made it through all right, though things looked badly for him as well." He also saved Dr. Tetzler, the brother-in-law of his friend Dirac in Cambridge. Dr. Tetzler, a Jewish merchant from Rumania, was thus spared from the German concentration camps, together with his wife. Heisenberg helped wherever it was possible; he did it unquestioningly and without making a great ado about it; he did not even tell me.

The Institute in Hechingen had become a refuge for many. Max von Laue was among those who had found protection and security at Heisenberg's Institute. In a certain sense, Heisenberg's exposed situation was like a shield protecting them all. And the problems of the individuals became more acute the closer the war came to an end, and the more the difficulties of life grew. During this period the people at the Institute were drawn closely together. They stood by each other through personal troubles, sickness, and severe mental stress to which they were nearly all subjected. A special treasure was the workshop of the Institute, where outstanding people were employed, who were always willing to come to the aid of the families to help deal with all their emergencies. Later, in Göttingen, they formed the basis of the workshop and helped to put together the new Institute with irreplaceable enthusiasm. They also took care to cultivate their relations with the townspeople — to some extent it was a question of survival to be on good terms with them. But as the people there were very friendly, this presented no problem. There were only a few convinced Nazis who were easy to avoid. Once Heisenberg and some of his colleagues even gave a small concert for the whole town, an occasion that was greeted with much gratitude and joy during the oppressive days of that dismal war. Thus, in a certain sense, it was truly possible to form "islands of stability, of freedom, and of trust," as Heisenberg had intended to do, and they succeeded in creating an atmosphere of humaneness and hope for survival among the members of the Institute and in its surroundings. No panic intruded into this atmosphere, in contrast to what was happening in so many other places, where people were either senselessly risking their lives or, filled with dread of what awaited them, were committing suicide.

A further duty Heisenberg felt bound to and that he thought to be important, was to give scientific lectures as often as possible, either at native or at foreign universities — especially, though, at the universities of the occupied areas, so as not to lose contact with his harried colleagues, but mainly to demonstrate that a different, better Germany existed than the Nazi Germany that had won the upper hand to such a terrifying degree.

For this reason, he accepted an invitation to Switzerland in November of 1942 that had been arranged by his close colleague Scherrer, together with the student body. To his astonishment he was even issued a travel permit by the German authorities. The agreement was that Heisenberg would give several scientific lectures in Zürich, Bern, and Basel in front of large audiences, and various colleagues extended friendly invitations, among them one to the home of Scherrer for a small dinner. Heisenberg accepted with pleasure. All he requested was that any political discussion be avoided during the meal and that only reliable members of the Institute be invited, so that he would not have to weigh his every word. Scherrer agreed to this, and the evening proceeded smoothly and without any disturbance. Later that night, a young man, whom he had noticed throughout the evening and whom he found exceptionally agreeable, accompanied him back to his hotel. On their way, their conversation was relaxed and animated. He told me about this encounter. Years later we received a book with the title *Moe Berg, Athlete, Scholar, Spy*. While leafing through the book, Heisenberg recognized Moe Berg as his young Swiss acquaintance who had accompanied him to the hotel, who had listened so attentively in the first row during his lecture, and who had participated in the discussion at Scherrer's with such intelligent and interested questions. At the time he did not know that the young man had a loaded pistol in his pocket, and that he had orders from the CIA to shoot Heisenberg on the spot if he detected even the slightest sign that Heisenberg was working on atomic bombs. In connection with this, Moe Berg writes that in spite of his psychological training and experience, he did not notice even the slightest twitch when he asked Heisenberg leading questions, nothing that would have given him the least cause to execute his order. This trip had another repercussion, as well, which fortunately also had a good ending. A strongly worded rebuke by the Gestapo was sent to Heisenberg's superiors, claiming he had made defeatist statements in Switzerland and demanding verification of the circumstances. It was his luck that this complaint was passed on to Professor Gerlach, who was a good friend of Heisenberg. Gerlach assumed a most severe and indignant air and gave his assurance that he would examine the case closely and hold Heisenberg responsible; it ended with a friendly discussion.

Naturally, this intrigue touched only on the very edge of our life. For one thing, I mention it because it, too, forms a small piece of the strange and variegated mosaic that I have undertaken to depict, and for another, because it shows how dangerous the situation surrounding Heisenberg became time and again. He was an individual with amazing luck in matters relating to his person. But: "Luck is a character trait" — that is an old saying, probably containing a good deal of truth. And perhaps, this "luck" really was a part of Heisenberg's character, for he combined many contradictions in himself, among them: prudent cleverness with naïve ingenuousness, a certain "peasant cunning" and a simple, straightforward, moral persuasive power, spontaneous comprehension of people and situations, and unflinching powers of endurance. And he confronted the anxiety caused by the danger of losing his autonomy and of no longer being able to make his own decisions with a deeply rooted and unbreakable resistance, so as to escape from this condition.

Perhaps this last trait of Heisenberg also contributed to his continuous refusal actively to participate in the conspiracy against Hitler, which peaked and foundered in an attempted assassination on July 20, 1944. It was not Heisenberg's nature to "go along" with something, if he could not influence its course. He trusted only himself; we have already encountered this facet of his character. Although he viewed the conspiracy with skepticism, he did know about it and was a close friend of several people involved, so that this part of the German tragedy also affected our lives deeply. And so, I will briefly describe the nature of Heisenberg's connection to the men involved in the attempt of July 20.

It must have been in the early summer of 1944 that Heisenberg received a visit in his Institute from a friend in the Youth Movement. It was Reichwein. He had belonged to the political wing of the Youth Movement that felt close to the socialist and pedagogical movement early on. Reichwein asked Heisenberg point-blank if he would be willing to participate in a conspiracy against Hitler. Heisenberg was horrified by Reichwein's careless indiscretion — he spoke in a loud voice and without any concern — and said to himself that if the enemy was being underestimated to this degree, the whole enterprise had to fail. He refused to participate. Then he emphatically warned Reichwein, and cautioned him to be more careful and close-mouthed if he wanted to achieve his object. But it was already too late for such warnings. A few weeks later, on July 4, Reichwein and some of his fellow conspirators were arrested and later executed for high treason. Heisenberg told me about this with great inner agitation.

Heisenberg was not a revolutionary, and in addition, he considered the time much too late for a revolutionary action. As has already been mentioned, he had placed great hopes in the military before the war, for he was of the opinion that a change of power could be forced only by someone with access to the means of power, and that was the military. But Hitler had very cleverly prevented that by strengthening the army and simultaneously permeating it with Nazi functionaries and officers devoted to him. At the outbreak of the war, Heisenberg saw another chance for military resistance; but the triumphal procession of the German army was so imposing that only a few retained the inner freedom and independence to still seriously think about a coup, let alone plan one. Thus, this chance was also squandered. A major part of the German military had been bought and blinded by the position of power it now inherited, as well as by its own successes. At the time, Heisenberg was deeply disappointed and embittered.

In Berlin — where, as already noted, he repeatedly had to travel from the more peaceful Hechingen, as part of the

Institute had remained there — he once again came into close contact with the circle of people planning a conspiracy against Hitler. Heisenberg had been asked to participate in the *Mittwochgesellschaft,* for decades an established conversation club of distinguished men from all branches of intellectual life, executive management, and the military. It rotated its meetings between the houses of members, and the host would then give a lecture in his own specialty, followed by a discussion. Then came the more sociable part. At the time, Ambassador von Hassel, the former Prussian Minister of Finance Popitz, Brigadier General Beck, the cultural philosopher Spranger, Sauerbruch, Diels, and others whose names I have forgotten, were members. Heisenberg was glad to take advantage of the opportunity to participate in this meeting. It was enjoyable and stimulating for him to get to know and converse with these different and interesting personalities. It gave him pleasure to learn something for himself from their experiences.

It must have been in the winter of '43-'44 that Popitz, who lived close to the home of my parents, asked him to stop by. During his visit, Heisenberg learned that a large coup was planned, and that thought was being given to the matter of how Germany should be better organized thereafter, when the Nazi regime had been removed and the war had ended through capitulation. Since Heisenberg himself constantly thought about this kind of question, an exceptionally fruitful and intensive discussion took place, creating a very trusting, albeit short, friendship.

To my knowledge, Heisenberg was never directly asked to participate in the conspiratorial activities by the *Mittwochgesellschaft,* in which most of its members were involved. In all probability, he would also not have been willing to do so, for he was mainly concerned with making himself available for the reconstruction following the war, and he thought that a revolution at that point hardly had a chance. He was of the opinion that, things having gone as far as they had, the war would have to be suffered through to its bitter conclusion. A revolutionary overthrow would no longer have been able to avert the catastrophe; perhaps it would have even intensified the confusion over right and wrong, so that the "afterwards" could not have been started with any clear guidelines, either.

On July 18, 1944, the *Mittwochgesellschaft* met once again in Heisenberg's Harnack House in Dahlem. Following this, Heisenberg wanted to travel to his Institute in Hechingen after having first spent the weekend in Urfeld with his family. On July 20, 1944, we heard about the assassination attempt on Hitler, about the failure of the coup, and of the death and the arrest of all the leading figures on the radio. The consequence was a wave of persecutions, arrests, and executions that went on for months and took the best of our people as its victims. Popitz, Beck, von Hassel, and many other friends of ours met their death. Spranger, who had not participated personally in the conspiracy, but had been involved only as a "contact person" like Heisenberg, was arrested as well. Heisenberg, however, was spared. Thus we really did survive the war and the Nazi *Reich*, and the American troops seemed to be saviors to us, for they brought the end of the horrible reign of terror and the war.

chapter seven

IMPRISONMENT AND THE DROPPING OF THE BOMB

On April 22, 1945, the French troops marched into Hechingen. There was no more resistance, just as Heisenberg and his friends had hoped and expected. A short time later, the French troops were followed by an American special unit, known as the "Alsos Commission."* They had been ordered to seek out, confiscate, and, if necessary, to remove everything in Germany having to do with nuclear research. This special commission was directly subordinated to the "Manhattan Project," itself charged with the development of an atomic bomb under the military supervision of General Groves, a large-scale organizer. Professor Samuel Goudsmit, a physicist, was assigned to "Alsos" as its scientific supervisor. Heisenberg had known him well for quite some time and had been in scientific communication with him in Michigan as late as 1939. The Alsos Commission had been able to find out that the center of the German "atom bomb production" was located in Hechingen and that the much sought after reactor of the Germans could be found in the caverns of Haigerloch.

Naturally, Heisenberg was not aware of any of this. He had reached an agreement with his closest colleagues that he would leave Hechingen heading east, when the hostilities seemed to have ceased, to stand by his family in Urfeld during the turmoils of the end of the war. He felt obligated to do this; after all, he had expected them to stay in the

---

*According to Armin Hermann, the name 'Alsos' is a Greek translation of 'grove.' "Alsos Commission" simply means: the commission of General Grove.

madhouse of Germany, although he could have spared them this fate. In the hope of retaining a small fund of uranium and heavy water to continue research on the reactor after scientific work had begun again, he had taken what remained of these materials and, together with his colleagues, he buried them in an open field. Prior to the dropping of the bomb, it was impossible to know that such hopes were unrealistic. Therefore, Heisenberg had agreed with his staff to reveal the hiding place only in case of an emergency. Thus they hoped that everything was under control, and that, in all probability, there would be no conflict situation demanding Heisenberg's personal presence.

This calculation did not prove to be false. All in all, the occupation of the Institute by the Alsos Commission took place without any problems, i.e., without any dramatic events. That the reactor in Haigerloch was already in possession of the Americans and being dismantled at the time was unknown to the physicists in Hechingen. The extent of the effort of the Alsos Commission was not yet known, nor the fact that they themselves, together with their reactor, were such a highly important war booty for the Americans. When asked where Heisenberg was, they freely answered that he was on the way to his family in Urfeld. And once the physicists in Hechingen had received the promise they could continue their work without disturbances, if they relinquished the uranium and the heavy water, the hiding place was also revealed. Later on, Colonel Calvert told me he had been under orders to threaten the bombing of Urfeld and, if necessary, to execute it, if the information had been refused. Luckily, it did not come to that. After having relinquished the uranium and the heavy water, however, something happened that no one had expected: the scientists suspected of being involved with nuclear fission and the construction of the reactor were arrested and transported to an unknown destination. From the Institute in Hechingen, they were: von Laue, von Weizsäcker, Wirtz, Korsching, and Bagge.

While all this was taking place in Hechingen, Heisenberg was riding east on his bicycle. He was on the road for three days and three nights until he arrived home safe and sound. Usually he travelled at night, so as not to be a target for low-flying aircrafts. During the day it was impossible to make progress anyway, and he spent most of the time pressed close to the ground in roadside ditches. I have already described his adventure with the SS man that could so easily have cost him his life, and which he survived only by bribing him with American cigarettes. But otherwise it must have been an extremely adventurous trip as well. He talked about it many times. Everything was beginning to disintegrate; he encountered bands of marauding, tattered figures speaking foreign languages, who had been released or had escaped from some prison camp or from forced labor, and who were now roaming through the countryside plundering; he saw groups of 14- and 15-year-old boys who had been drafted and abducted the last minute, and who were now camping along the side of the road, crying, hungry, and lost, not knowing what to do; he met hordes of soldiers of the most varying nationalities, going some-where, some to the east, others to the west or north, without a plan, exhausted, and threatening. It was always neces-sary to hide and make detours. The town of Weilheim, where he went in hope of perhaps finding a train that would take him a little farther — for he, too, was starving and exhausted to the limit of his endurance — was in flames when he arrived there. He slept on his bicycle in the half-destroyed train station for a few hours, before he could continue on his way — and then, suddenly, there was a train that took him along for a couple of kilometers, after all.

And finally, suddenly and unexpectedly, I saw him com-ing up the mountain, dirty, dead tired, and happy. I will not indulge in a description of the last days of the war; but we all fell into an ecstasy of relief on the evening we heard the message of Hitler's death on the radio. Now it could not last much longer! We fetched the last bottle of wine — we had actually been saving it for the baptism of our daughter — from the cellar, and drank it with tears of relief and deliverance. There was no thought of going to sleep; hope, once again, let the future blossom before our minds' eye.

Two or three days later, five or six heavily armed soldiers appeared on our terrace. I was in the house alone. Heisenberg had walked down into the village to look after his mother, who, due to the lack of space, could no longer live in our cottage, and therefore now lived in a room on the lake. Seeing these heavily armed soldiers approaching, I was very frightened, since the SS had been spreading terror in our region for days. They hanged soldiers who had been released from a convalescent company on trees for "desertion," or set fire to farmhouses, because, being scared, the women had hung out white flags. I was most relieved when I realized it was not the SS. In contrast, the American soldiers looked like liberators from a horrid nightmare. Colonel Pash, the Russian-born trooper leading the small special unit, instructed me to call up Heisenberg and to tell him to come home, without telling him why. Heisenberg, who had already seen the American tanks enter Urfeld, understood the situation immediately and came up without delay. This was followed by long interrogations. At one point there was some shooting down in the village, filling the Americans with apprehension, since they themselves were so small a group, and they were not aware that it was only an individual action. At night our house was guarded; we could hear the steps of the soldiers walking around our cottage. The next day, Heisenberg was taken away by an escort of tanks and armored reconnaissance vehicles that had arrived in the meantime. We were not informed where he was being taken. All we were told was that he would be back in three weeks. The three weeks turned into eight months; for us, these months meant torturous uncertainty and a hard struggle for survival.

During these eight months I received three short, terse reports, indicating that he was still alive and doing all right. Since they were delivered by an American officer, I assumed that Heisenberg was in America. In reality, though, he was interned in England together with the other nuclear scientists who had been arrested in Hechingen; they had been joined by Hahn, Gerlach, and Harteck from Hamburg, and by Dr. Diebner. Even the motives of the arrests were a puzzle to us, and we were unable to find out anything about them. Later we were told that the leading nuclear scientist had been taken into custody to prevent their being captured by the Russians, who possibly could have obtained important scientific or technical information on the construction of reactors or, and this was the main reason, the atom bomb.

Heisenberg was first taken to Heidelberg. It was one of the few large cities that had not been destroyed by bombs, and thus it offered the Americans the possibility of requisitioning sufficient living and administrative space. From Heidelberg, I received one last detailed letter. He was only allowed to write about how he was feeling, and about nothing else. In this letter he wrote about the spring, already in its full splendor in Heidelberg, and about the nice house where he was quartered, and that he was finally able to eat to his heart's content again. "...in my emotions, the misery of the past and the sight of the human destruction are mixed with the intensive happiness of making a fresh start and building anew." And farther down, the sentence: "I hope that fate will allow me to be equal to my task."

In Heidelberg, Heisenberg met with Goudsmit. Goudsmit, originally a Dutchman and now an American, was to interrogate Heisenberg, too, and his objective was to find out how much information the Germans had on the construction of an atomic bomb. To the great astonishment of the American scientists, they as yet had not found anything other than the little reactor in Haigerloch, and no trace of an atomic bomb. And they were encountering scientists who were by no means trying to hide their knowledge from them; on the contrary, after a short time, they were willing to reveal what they knew and talked about it. This appeared wholly unbelievable to the American commission. Had they not constantly lived in the belief that the Germans had gone to great lengths to construct atom bombs? And had they not constantly feared that the Germans were possibly ahead of them in the development of atom bombs?

But to Heisenberg everything appeared quite different. On the very same evening of the interrogation he wrote to me: "The conversations with Goudsmit and Kemble were as amicable as though the last six years had never taken place, and I myself haven't felt this well for years, both emotionally and physically. I am full of hope and ambition for the future. Naturally, there will be setbacks again, but there is no reason to let that be distracting."* For him, it was a return to the free world to which he felt he belonged. So, during the course of the long discussions, he asked Goudsmit what was going on with the atomic bomb in America, whether work was being done on it. With a smile, Goudsmit answered that there had been more important things to do during the war, and that there had been no efforts in that direction. This information was complemented by the fact that German intelligence had never supplied any concrete reports on construction plans that would have been evidence of such a large undertaking. The vague threats that had reach Göering via Lisbon in 1944, that Dresden would be destroyed by an atomic bomb if Germany did not capitulate within six weeks, had, after all, not been true. Dresden had not been destroyed by an atomic bomb. So what could have been more natural than to believe Goudsmit's statement? And Heisenberg believed him without reservations.

Perhaps there was also a deeper psychological reason for

*This letter, as well, I found in his posthumous papers; he was no longer allowed to send it to me at the time he wrote it.

19 In the winter of 1946/47: Elisabeth Heisenberg and Max Planck at an invitation of the English in Göttingen (German Scientific Advisory Council)

20 Heisenberg with his new seminar in Göttingen in 1947: Schlüter, unknown, Heisenberg, unknown, Koppe, and Wildermuth

21 Elisabeth Heisenberg in 1952

22 Heisenberg shortly after the war (around 1946)

23 Wolfgang Pauli, Zurich 1953

24 Karl Wirtz

25 Siegfried Balke (at that time minister for scientific research), Otto
Hahn, Konrad Adenauer, Werner Heisenberg, and prime minister
Hellwege during the ninth general assembly of the Max Planck Socie-
ty in Hannover in 1958

Heisenberg's credulity: in Copenhagen, he had made such great efforts to achieve just what Goudsmit was telling him now and he was all too willing to believe it; it was such a pleasing thought, that his mission might have made a small contribution to preventing the disaster of this development. On the other hand, Heisenberg was naturally aware that Goudsmit was sworn to deepest secrecy and not permitted to say anything about the bomb. But for Heisenberg, the war was over; he was a prisoner of the Americans, and he was now irrelevant in the further course of events. Had Goudsmit said that he could and would not answer these questions — after all, the war was still going on — Heisenberg would have accepted this and it would have been different. But he had quite decisively said: "We made no efforts in that direction; we had more important things to do." It made sense to Heisenberg. Thus, he felt betrayed by Goudsmit, and when, after the bomb had been dropped, his demeanor was interpreted as arrogance and absolute ignorance, he once said with some bitterness: "How was I to know that Goudsmit was lying right to my face."

In this case it is probably also better to examine the situation more closely, if justice is to be done to both men, meeting each other again with completely different positions. First, it has to be said that Heisenberg was probably wrong with his accusation that Goudsmit intentionally lied to him. One may assume that Goudsmit in fact possessed little or no information on the building of the atom bomb. Specifically this lack of information is the probable reason for his having been chosen for the mission. Since the main concern of this whole action was to prevent the Russians from obtaining any information on the atomic bomb, it was important that the responsible American authority had no information on the state of affairs, as he himself could fall into the hands of the Russians. On the other hand, it is improbable that he knew nothing at all; in any case, to Heisenberg, it looked as though Goudsmit had been consciously deceiving him, and that he had let himself be deceived. That hurt him.

The course of the remainder of the long conversation was not much different. Neither of the two men, who formerly had been bound by a collegial friendship, was able or possessed enough imagination to guess what the other was thinking. In reality, there was never any genuine contact. Nevertheless, the old, well-practiced forms of friendship still functioned and hid the divisions separating them, and Heisenberg was a victim of the illusion. Once again he was trapped in his old dream that some of the former openness and trust, that had existed in the "international family of physicists" before the war, was still present. He thought that it would all reappear if only they faced each other again, and he could and would not believe that events had already banned him from this trust long ago, that the reality of politics has far greater powers than any ideational commitment. He was not the only one who had fallen victim to this error. His letters written to me after his internment make it clear how much he hoped things could now return to the way they had been. The optimistic side of his nature had once again won the upper hand; so he was glad to believe that Goudsmit trusted him. And so, this positive and endearing side of his nature led him astray again.

In contrast, Goudsmit had everything but trust in Heisenberg. He distrusted him deeply. For him, Heisenberg was a German who had demonstrated his agreement with what had happened in Germany by refusing all the offers from America and by staying in Germany. He, Goudsmit, had been standing in front of the destroyed and plundered house of his parents in The Hague just a few short days before. He had grown up there. And he had recalled the last letter from his parents before the Nazi henchmen had arrested them and transported them to one of those terrible extermination camps in a freigh car, like "cattle," and had "liquidated" them in a gas chamber. He was still filled with inner turmoil, through feelings of grief, horror, guilt toward his parents, and, naturally, hatred. The whole terrible truth was just revealing itself to him now, as well. In spite of this, he had spontaneously greeted Heisenberg with the question: "Wouldn't you want to come to America now and work with us?" But when Heisenberg answered: "No, I don't want to leave, Germany needs me!" he could only interpret it as meaning that Heisenberg was totally blinded by his own importance and achievement; and he was convinced that Heisenberg's actions were a combination of opportunism, arrogance, and presumptuousness. There was no more common ground between them.

After the completion of his mission in Germany, Goudsmit returned to the U.S.A. and wrote a book about his experience as a scientific expert with Alsos. It was entitled *Alsos* and was published in 1947. Certainly, there was a strong need to find out about the potential development of the atom bomb within Germany, the Germany that not so long ago had caused such great anxiety, and that was now so pitifully broken. And no one seemed more suited to report on it than Goudsmit; he knew about it first-hand. In fact, the Americans had developed extremely exaggerated notions about the scientific and technical possibilities of this basically small nation, and the time seemed right to set things straight. Obviously, however, it was too soon, and emotions were still running too high to make an objective and just report. Goudsmit's portrait of German atomic physicists, but specifically of Heisenberg, hardly coincides anywhere with reality, and almost sounds like a foul caricature, so that, in response, American newspapers printed large, bold-faced headlines: "TOP-NAZI Heisenberg."

Heisenberg exchanged a few letters with Goudsmit about the latter's book, and it is characteristic that he leaves aside any justification of his own personal actions. He only refutes the notion that the responsible German scientists had known nothing about constructing a bomb. In his letters of January

5 and November 3, 1948, addressed to Goudsmit, Heisenberg responds that they had been quite familiar with fast neutron chain reactions, and he continues: "Aside from that, I do not find this fact important because I believe the results of our work to be special, scientific achievements of which we ought to be proud; to the contrary, I believe that this whole development was practically unavoidable following Hahn's discovery and the work of Bohr and Wheeler. In my opinion, the great achievement of the American and English physicists is mainly to be seen in the incredible effectiveness of the technical execution, in the systematic use of the vast resources that could only be made available by the gigantic industrial potential of America."

Goudsmit later regretted having written the book, and apologized to Heisenberg for it; nevertheless, the book is one of the reasons for Heisenberg's character falling into such ill repute. Heisenberg had long resigned himself to the fact: "You know," he said to me, "history is always written by the victor; you just have to live with that." But this implies that history is viewed the way the victor saw it or wanted it to be seen, and that is not necessarily what really happened.

Heisenberg was not released following the interrogation in Heidelberg, as might have been expected if one believed the promises of Colonel Pash. Instead, he was driven to Paris where, to his great surprise, he met the other nuclear physicists who had been interned. There were nine scientists now: von Laue, Hahn, Gerlach, von Weizsäcker, Wirtz, Bagge, Diebner, Korsching, and Heisenberg; a few days later, Harteck from Hamburg was added to make it ten. Their lodgings in Paris were extremely spartan and they were under strict military guard; they were treated like war criminals. But this condition did not last long, and after a temporary stay of a few weeks in a small country castle in Belgium the group was finally flown to England, where they were quartered in an old country estate called Farm Hall, located near Cambridge and abandoned at the time.

This was a thoroughly comfortable lodging; the well furnished house was situated in a large garden with roses growing wild and wide expanses of lawn. The house and the kitchen were tended by German prisoners, among them a cook. A friendly English officer took care of them. The ten German scientists lived in Farm Hall as though they were in a golden cage, and they wanted for nothing. Heisenberg was even given a piano, and although he lacked the sheet music to play from, he enjoyed recalling the great pieces of music he had once learned by heart, and playing them over. In addition, they had a library with all kinds of good English literature; they could work in the garden; and in the evenings they played bridge. They resumed work on their sciences, especially the theoreticians for whom it was easier, and became immersed in uncontroversial, fundamental problems. There were daily seminars with scientific discussions. This life lasted half a year. The news of the dropping of the bomb on Hiroshima burst into the midst of it.

The Americans had installed listening devices in the rooms of the prisoners, with which they taped all of the conversations of the German scientists. Used to this method from the Nazis, the Germans speculated on the possibility. Some of them started to look for the "bugs" behind the pictures and under the carpets. Others reacted with more equanimity. Later, when we met again, Heisenberg told me about it and said that, as a joke, he had wanted to make fun of the Americans or English, and in case they could hear it had said: "It is hard to believe that they would use these Gestapo methods in 'merry old England!'" With that the matter was closed for him, and he no longer worried about being heard or not.

The tapes made at the time have been kept under lock and key in England or America since then. They are frequently quoted, but to my knowledge, they have never been read by anyone having a full command of the nuances of German and who knew the people speaking, thus being in a position precisely to interpret the discernible discussions. General Groves, the organizational head of the "Manhattan Project," published English translations of excerpts from these tapes in his memoirs *Now It Can Be Told.* Moreover, he continuously refers to them in his chapter on the German atomic scientists without necessarily quoting the relevant passages. The German translator of these memoirs, who is quite aware of the problem of translating the tapes, adds a footnote: "The author went to much trouble to obtain the original text of both this recording and the one appearing later in Chapter 24, but was unable to find a trace of either. He thinks it is possible that at the time only the English text was sent to Washington. The German text presented here is thus a re-translation from the English."

Experience now tells us that the original meaning of any text can be distorted quite easily through a double translation. A double translation remaining true to the original meaning is hardly possible, especially when dealing with such emotional texts as permeated with politics and difficult psychology as these are. This explains why most of the texts sound strange and distorted to me. And after all, I knew well all of the people speaking. In the English translation, the above-mentioned passage regarding the built-in listening devices takes on a completely different coloring, as well. In the book by General Groves it looks like this: "Microphone installed? [Laughing] Oh, no! they're not as cute as all that. I don't think they know the real Gestapo methods; they are a bit old-fashioned in that respect." Well, I do not know Heisenberg's original wording. What is certain is that this translation reflects nothing of what he told me: nothing of his small, cute joke. In the translation back into German, also used by Armin Hermann in his biography of Heisenberg, it sounds even harsher and more arrogant: "Mikrophone eingebaut? [Lachen] Oh nein! So gerissen sind die nicht. Ich glaube nicht, dass sie die wahren Gestapo-Methoden kennen; in dieser Beziehung sind sie ein bisschen altmodisch." I just cannot believe that Heisenberg could have spoken in this way.

Since these tapes are quoted so frequently in the Unitd States and are used as proof for the presumptuousness and lack of knowledge of the German scientists, the objection should be made that the interpretation of such documents as these tapes must be wrong of necessity, if the originals are not examined by experts, familiar with the nuances of the German language; they must also be capable of including correctly the psychology of the various individuals in their analysis, especially the complicated relationships among them. Thus, I would like to probe this psychological situation somewhat more deeply.

Naturally, and for varying reasons, the dropping of the atomic bomb was an enormous shock for the ten scientists, different as they were from each other. Each one reacted differently in this situation, but all of them were shaken and deeply confounded. They knew the terror of air attacks all too well, and every one of them had imagined the destructive force of such a bomb too frequently to be able to evade the horror. There was naturally also some admiration involved, and some of them perhaps even felt some envy and bitterness. Hahn's reaction, by contrast, was one of despair. He felt guilty, and his friends were seriously concerned he might harm himself. Heisenberg reacted with doubt and disbelief. That was not atypical of him. He was always incapable of believing right off in the truth of something horrible. So now he just refused to accept it; he called it a bluff and devised all sorts of untenable theories as to what type of bomb it could be. His deeply rooted, defensive stance against the horrible was certainly enhanced by the fact that he had imagined the technical and organizational problems of the construction of a bomb to be so large all along that he had thought it completely inconceivable that the Americans could really have succeeded in mastering all these problems in so short a time. He had probably also underestimated the human and technical resources of the Americans. And indeed, it was an incredible achievement! But he also remembered Goudsmit's statement, that the U.S.A. was not involved in an atom bomb project. This was certainly also a reason for his puzzling over the bomb for so long, and why he devised incorrect hypotheses. He did this until the facts would finally permit no other interpretation.

There can be no doubt that the tapes would aggravatingly reveal the human, all too human, aspect of this group, forced together almost by chance. In no way were the ten scientists a homogeneous gathering. They were almost all strong individuals, though by no means were all of them self-possessed and aloof, and they held very different points of view in human as well as in political affairs. Each reacted to the internment in his own way; some of them were very disturbed and irritated by it, and developed strange reactions and sensitivities; others were worried about their families who had been left behind in Germany in the unsettled period at the end of the war, and from whom they hardly heard; others thought their internment to be outrageous, while some mastered the situation more gracefully with equanimity and humor. But they were separated even more strongly by their past, for it was precisely in their political origins that they differed most severely and where they could find no common ground. There was the one who had played an active role in the Party and who had been superimposed on them all as a control, or the one who still thought in categories of Nazism or at least nationalism, and who was full of suspicion and resentment toward those who did not. There was also one who, fully implacable, had retreated into passive resistance, or the one who, like Heisenberg, had tried to continue the fight against National Socialism, at least in his own arena, and to preserve and do what was right and possible in it. They all held a different opinion on the war as well as on the success or the failure of their common effort. The scale of opinions was broad, and the differences below the surface were sharper than what met the eye. True, they were all practiced in objective thinking, but what had occurred had not entered into the realm of objectivity by a long shot. To the contrary! The past was stuck in all of their throats like a bitter lump that they could not simply swallow, for, one way or the other, they had all become guilty, just as he, who had participated in the events in any manner, or he who had left the stage and had let them take their course, was guilty, or those who had participated in the construction and the dropping of the bomb. They had all become guilty, but they were not all aware of it with the same intensity, and this was the cause of great tensions among them. So as to enable them to live together, talk of politics had to be kept to a minimum, and with the exception of small, intimate circles, discussions were limited to everyday events or science. General Groves notes this with great surprise and complete incomprehension.

I would like to use the example of Otto Hahn to show how psychologically difficult the interpretation of the tapes is. Hahn had a special position in the circle of the ten German scientists in Farm Hall. Due to his magnanimous and endearing nature he was the mediator between the individual opponents, and he was able to smooth out the serious differences and tensions that kept arising. But Hahn was not initiated into the problems of the construction of the atom bomb. During the war he had worked on chemical problems of a completely different sort, and he was not part of the circle of Heisenberg's colleagues in Hechingen. Hahn's institute was in Tailfingen; they did not see each other often, and topics as explosive as these were not discussed, either then, or now in internment. In addition, there was a second factor: Hahn had the gift of a very specific type of humor that he would always use when the situation was especially tense and charged, or if he himself was completely depressed. It was a type of gallows humor, and later on he once said to me: "Oh sure, I always played the clown, but actually my heart was weeping the whole time!"

But as a result, he was remarkably capable of defusing and easing a desperate situation. In order to understand Hahn's remark, this must be known. In English it would read as follows: "If the Americans have an uranium bomb then you all are second-raters. Poor old Heisenberg!" And later, again: "At any rate, Heisenberg, you're just second-raters, and you may as well pack up!" The English words do not reflect what he actually sounded like, but still, this is his style of making a joke by using a few somewhat aggressive words, and thereby freeing the situation of its dangerous edge and tension.

A small episode may serve to illustrate this. During the first large-scale air attack on Berlin, all of the leading scientists were sitting together in the air-raid shelter of the Ministry for Aviation, since, on Göring's order, they had to attend a lecture on the effects of high-explosive bombs given by a physiologist. The speaker had given a lengthy explanation that death by high-explosive bombs was actually a humane death, as the air bubbles of the lungs would burst immediately due to the excess pressure, and the person would be dead on the spot. When the large high-explosive bombs came whistling down and exploded in an inferno of terrible noise, and everybody, including the lecurer, groaned aloud in the dark cellar, Hahn's voice suddenly rang out: "This guy, Mr. So-and-so, doesn't believe his own theory anymore!" The effect of Hahn's words was incredible; Heisenberg told me about it full of admiration. The terrible tension relaxed, people laughed with relief, and the spell was broken.

In many places, General Grove's book does the German scientists more justice than the one written by Goudsmit. The description of the events in Urfeld, however, is only remotely related to reality. And along with a host of other contradictions, the conclusion he reaches on the work of the German nuclear scientists bears little semblance to reality. He writes: "The Germans had not thought of using the bomb designs we used. Ours took advantage of fast neutrons; the Germans thought that they had to moderate them as in a pile. In effect, they thought that they would have to drop a whole reactor, and to achieve a reasonable weight they would need this enormous amount of Uranium-235." This is contradicted by the certifiable statements made by Heisenberg during the meeting of June 4, 1942, where, in response to the question of how big a bomb would have to be to destroy a city like London, he answered: "As big as a pineapple." But to accept this state of the knowledge of the Germans, General Groves would have been forced to discard the idea that the Germans knew nothing about the construction of the bomb but had been working on its development with all their strength. However, he was not capable of making this concession.

To conclude this chapter, I would like to attach the memorandum written by the German scientists a few days after the bomb had been dropped. It reads as follows:

"Since the press reports of the last few days contain partially incorrect representations of alleged work in Germany on the atom bomb, we would like briefly to explain the development of work on the atom bomb here.

1. The fission of the nucleus of the uranium atom was discovered by Hahn and Strassman in December of 1938 at the Kaiser Wilhelm Institute for Chemistry. It was the result of purely scientific research, which had nothing to do with a practical application. That nuclear fission brought about a chain reaction of the atom nucleus, and thus its technical exploitation for powering machines, was almost simultaneously discovered in various countries only after publication.

2. At the beginning of the war, a research group was created with the task of examining the practical application of this energy. By the end of 1941, the initial scientific work had proved that it would be possible to use nuclear energy for the production of heat and thus to power machines. On the other hand, it did not seem feasible to build an atom bomb at the time of the existing possibilities

in Germany. Thus further work was concentrated on the problem of the machine, requiring heavy water as well as uranium.

3. To this end the plant of Nors Hydro near Rjukan was expanded for the production of large quantities of heavy water. The attacks on the plant, at first by the commando mission and later by airplanes, halted this production.

4. At the same time attempts were made in Freiburg, and later in Celle, to dispense with the need for heavy water by concentrating the rare isotope U-235.

5. The experiments on energy production were continued at first in Berlin, and later in Haigerloch [Württemberg] with the remaining reserve of heavy water. Toward the end of the war this work had progressed to the point where the construction of an energy producing machine would probably only have taken a short while longer."

This memorandum contains nothing about the actual problems that the physicists were confronted with during the war. But at the time it was much too early to say anything about them.

The ten prisoners waited for their release from day to day. They were not subjected to any normal jurisdiction; they had not even been declared "guilty"; rather, they were being held "in custody" as virtual war booty that was not to be in the possession of either the Russians or the French. Their official status was: "Detained at his Majesty's pleasure," a curious form of imprisonment. And the prisoners frequently asked their officer on guard "whether his Majesty had not yet derived sufficient pleasure from their detainment!" But in spite of an absence of jurisdiction, this type of imprisonment has its firm rules: it is not allowed to last longer than six months; and thus it happened that the ten prisoners were brought back to Germany exactly six months later to the day. At first, they were still held in a kind of half custody in Alswede, a small town close to Minden, which was the English headquarters at the time. There we women were finally allowed to see our husbands again.

Then, after several weeks, some of them were released. Heisenberg and Hahn had to stay. For there was still uncertainty where Heisenberg's Institute and the general administration of the Kaiser Wilhelm Society were to be located. Nevertheless, the new beginning was tangibly close, the new life we had been hoping for throughout the dark times.

chapter eight

AFTER THE WAR

In a certain sense, time stood still for the scientists in Farm Hall during the time of their internment. On the outside, however, great decisions were being made; the world was changing. What was to become of the Kaiser Wilhelm Society? Its president, Dr. Vögler, had been killed in Dortmund at the end of the war. And the remark "In dissolution" was entered beside the name Kaiser Wilhelm Society in the public records of the county court in Berlin. Thus there was a strong movement simply to disband the Kaiser Wilhelm Society. But there were also other voices within the occupational forces and, on the German side, the Society had an unusually capable Secretary General in Dr. Telschow. Counter to the orders of the government, he had moved part of the general administration from Berlin to Göttingen in anticipation of Germany's defeat; in addition, he had freed a considerable sum of money, a "last reserve," and had distributed it among the various institutes, so that they could survive the coming great crisis.

But most of the directors of the Kaiser Wilhelm Society were unavailable for the group in Göttingen, in part because they were in England, like Hahn and Heisenberg, or because they were being held in "temporary custody" as "very important persons," as it was then called, or simply because the institutes were scattered throughout Germany and were thus located in different occupational zones. At the time, the borders of the zones presented an almost insurmountable barrier. Given these circumstances, Telschow decided to act on his own initiative.

It was very fortunate for the Society that Max Planck arrived in Göttingen as a refugee on June 4, 1945. Indeed, he was an old and broken man. His son Erwin had been murdered by the Nazis at the end of the war despite all interventions by Planck and his friends. This had a deep and inconsolable effect on him. He himself had just gone through the harrowing experience of a refugee. He had found protection from the bombing of Berlin through the kindness of friends near Magdeburg; but at the end of the war this region as well was entangled in combat action. The Plancks had to move on once again. The old scholar and his wife then wandered through the woods, suffering pains, dangers, and all imaginable hardship, until the war had finally moved on again. When they returned to their property, they found everything completely destroyed. But they stayed there until an American jeep drove by one day and picked up the suffering, old man and took him to Göttingen, where he was taken in by loving relatives. That, too, was executed by Alsos; they wanted to prevent Planck from falling into the hands of the Russians, if the zone he was in were to come under their control.

Nevertheless, Planck's appearance in Göttingen was a great help to Telschow. Before Dr. Vögler, he had been the President of the Society for eight years. And now Telschow asked him to take over the Presidency on a provisional basis once again, and to help with the new structuring of the Kaiser Wilhelm Society. In spite of his weakness, Planck did not shirk this responsibility; and probably, in the time directly following the war, his acceptance meant the salvation of the Society. Now, once again, it had a center from which negotiations could be conducted and undertaken.

At first, the most important thing was to determine what was left of the many institutes, to apprise the various staff members of the continuation of the Society, and to encourage them to persevere. Whoever remembers the difficulty of such trips in the still absolutely chaotic conditions of the time will be able to appreciate the actually heroic decision of Telschow. Mrs. Bollmann, Telschow's faithful travel companion and his longstanding assistant, reports as follows: "Thankfully, the British Military Government in Göttingen made a confiscated Mercedes Benz diesel available to us that took us. . .to the institutes in all three zones during the coming months. . .Every trip was an adventure. Identification controls, searches for provisions, examinations, continuous breakdowns, lack of fuel, missing replacement parts, great difficulties during the exchanging of respective food stamps, worries about lodging, etc., alternated in colorful sequence. . ." * But Telschow's efforts paid off. What he found was most encouraging. In spite of all the destruction, the Society still had a considerable foundation of functional institutes, in which scientific work could be resumed at least to a small degree.

On July 24, 1945, Planck then sent a circular letter to all of the directors who could be reached and to all of the gentlemen of the administrative board who could be found, in which he proposed electing Otto Hahn as the new director of the Kaiser Wilhelm Society. Otto Hahn still possessed an unbroken potential of trust at home as well as abroad. He was conciliatory, reliable, and universally liked; without a doubt, he was the right man. The suggestion was immediately adopted unanimously, but, for the time being, Hahn was still ineterned in England.

This was roughly the condition of the Society when the ten internees returned to Germany in January of 1946, and were gradually allowed to return to their families and their places of work. Harteck was released to Hamburg, and Gerlach, who was not permitted into the American zone, to Bonn. Heisenberg did not receive permission to enter the American occupational zone either, and thus could not yet visit us in Urfeld. To his great disappointment he was not allowed to travel into the French zone to reclaim control of his Institute. The chief supervisory council had decided

---

*Erika Bollmann, *Erinnerungen und Tatsachen*, Stuttgart, 1956.

that Heisenberg's Institute would be relocated to the English zone together with the general administration. Hamburg and Göttingen were being discussed for this. As early as January 25, 1946, Heisenberg wrote to me from Alswede: "It now looks as if we might be located in Göttingen, but the decision rests with the highest authorities...," and the highest authorities could in no way reach an agreement, so that a decision was a long time in coming. But finally the die was cast for Göttingen. It had not been destroyed, and it offered enough space for the new institutes and the administration of the Society in the old halls and buildings of the aerodynamics test plant, "the AVA." Moreover, Göttingen was invested with a long and glorious tradition in the natural sciences that was also highly respected by the English. This could be utilized.

On April 1, 1946, then, Otto Hahn took charge of the affairs of the Kaiser Wilhelm Society. But this, too, was only a temporary arrangement, and there were still countless, tenacious negotiations prior to the foundation meeting of the new Society in September of 1946 in Bad Driburg, where Hahn was officially proclaimed as the new President. One of the first official acts of the new Board of Directors was to rename the society, for which a new, more democratic order was planned, the "Max Planck Society." Max Planck, as well, was a person in whom trust was still complete, and he had been the President of the Society for years.

Since he had so consciously kept himself ready for the time of reconstruction, the inactivity he was subjected to by his internment in England had been hard for Heisenberg to bear. There he was, in a golden cage in England. The only thing he had been able to contribute to a favorable development from there had been a few discussions he had had with some of his English colleagues during two meetings of the Royal Society in London. They had been friendly enough to invite Hahn and Heisenberg to these meetings of the Society, of which the two were members, toward the end of the year. On this occasion, Heisenberg mainly spoke with Professor Blackett, whose political influence in England was large at the time. Blackett then actually did take a decisive role in the tenacious negotiations surrounding the reconstruction of the Kaiser Wilhelm Society and the rejuvenation of German research, arguing the German side. It is surely owing to him that an English commission, led by Dr. Bertie Blount, was assigned to the German team in Göttingen, with the task of supporting the efforts of the Germans, but also to make sure that the newly created institutes only got involved in basic research, and not in anything having to do with nuclear physics, which could lead to the construction of reactors or even atomic bombs. A very fair and effective cooperation arose between Colonel Blount and his group, on the one side, and the leading scientists, who in time gathered in Göttingen, on the other; especially between Heisenberg and Blount, this developed into a friendship that lasted for decades.

Reading this short report on the reconstruction of the Kaiser Wilhelm, i.e., Max Planck Society, it is hard to imagine the difficulties involved in each step forward. The chief supervisory council had to give its consent to nearly every one of these steps; and there was by no means agreement between the four powers, Russia, France, England, and America, neither on the method of their approach, nor on their objectives. Also, there was a strong tendency to keep the jurisdiction of the Germans as small as possible. This situation introduced a strong element of uncertainty into all of the decisions of the Germans. Besides, the mobility of the individuals was extremely limited; obtaining passes was a tiring and time-consuming affair. The few cars that existed were fueled with wood, for there was no legal way for an average mortal to obtain gasoline, and the trains were generally still in a miserable condition. I myself once travelled from Urfeld to Frankfurt at the beginning of

November, to get some information on the fate of my husband in the American headquarters, but also because I was running out of money, and I no longer knew how to pay in future for the little we were just barely getting by on, as it was. This trip took me three days and two nights, without a warm meal or any lodgings; the trains were dirty, crowded, without windowpanes. There were no rooms in Frankfurt either, since it had been largely destroyed. Finally, as though by a miracle, I was standing in front of Professor Robinson, the man at the American headquarters who had to know about my husband. I was so starved, dirty, and exhausted that he took me to his apartment straight away and gave me something to eat; then he arranged for a room where I could sleep. I learned that my husband was doing all right and that I need not worry about him. And then he had an English Colonel take me to Hechingen, where I was again supplied with the necessary money for a few more months.

That was in November of 1945; by 1946 the conditions had already grown slightly better. Nevertheless, every step forward was still a tiresome struggle. At times, almost everything appeared to be hopeless, and some of our acquaintances gave up and looked for ways of going to America, which looked like the Eldorado of life to many at the time. Heisenberg also still received inquiries as to whether he would like to emigrate to America. And he still refused without hesitation. He had made his decision a long time ago; he had already written to me from Alswede: "You ask me whether we had the choice of staying in Germany or going to America. Hahn and I were asked semi-officially. Goudsmit already asked me whether I want to go to America during my first 'interrogation' in Heidelberg, and Blackett raised the question again in England. I had already given the matter a lot of thought beforehand and roughly took the following point of view: I am aware that America will be the center of scientific life during the coming decades, and that the conditions for my work will be much worse in Germany than over there...I, for my part, want to try and help with the reconstruction here during the next few years, and if the discord of the politicians is not too disturbing, it should be possible to awaken something of the lively intellectual life of the 20's...That it would, in many ways, be nicer and more comfortable to live in America is something one just has to accept." But his perseverance and his steadfastness were once again severely tested.

   While the German working team, consisting of Heisenberg, von Weizsäcker, Wirtz, and Bagge were planning the new beginning of the Institute, the large wind tunnels of the AVA were being dismantled and transported to England, and all of the scientific machinery and commodities, with which they had secretly hoped to be able to start anew, were being carried away, so that basically all that remained were the empty rooms. But it was impossible to reach Heisenberg's Institute in Hechingen in the meantime, as it was located in the French occupational zone. Bringing the people, equipment, apparatus, and the families with their household items to Göttingen was an undertaking involving untold weary negotiations, setbacks just when there was hope of having advanced a step, and new negotiations; the whole thing took months. We, the family, lived in the American zone, and it was almost impossible to obtain travel papers, especially for nuclear researchers who, as Heisenberg once wrote somewhat bitterly in a letter, "seem to be viewed as particularly dangerous people." Blount and his group helped where they could. But they were powerless against the decisions of the control board and the complicated conditions.

   All this, and the poor condition of his health caused by the period of hunger, to which his body had great difficulties adjusting following the "fat days of internment," brought Heisenberg to the limits of his physical and psychic endurance. In 1946, on the Sunday of Pentecost, he wrote me a letter: "Well, I'm not doing so terribly well. . .I don't really know what is the matter; I'm certainly completely healthy, but I'm always totally exhausted. . .but it is not just the hunger. I am no longer equal to this continuous organizing, with its many disappointments. It would be very important not to be without companionship all the time. But for the time being Göttingen is completely dead for me." He had not thought the beginning would be so hard, when he had left Urfeld a year ago with the hope that a new life was to commence.

After having received this letter, I decided to join him in Göttingen, whatever the price. Fortunately, a very capable and reliable young girl, to whom I could entrust the children for some time, had joined my household during the chaos of the end of the war. And so I set out. This trip, as well, still was long and tiring, in addition to being exciting, since I did not possess any travel papers. But I was lucky and got there, arriving in Göttingen at four in the morning, worn out and freezing cold. With my rucksack shouldered I walked through the sleeping streets of the pretty, little town over to the AVA. I rang the bell of the tightly locked door, and a sleepy porter looked out of the window and irritably asked me what I wanted. I was Professor Heisenberg's wife, and I would like to go in and see him. Well then, he retorted, did I have an English pass for the AVA. Naturally, I did not; Heisenberg did not even know I was coming. But a couple of minutes later he was standing in front of me. I was still not allowed to go in, but he came out, took me into his arms, and then we walked up the Hain Mountain. It was a clear morning in early summer. The birds were singing their lovesongs; the forest was decked out in tender green; the air was clear and pure. All of the fatigue and chill of the night had been blown away. Heisenberg knew about a bench in the Kerstingröder Field; we went there. It was overgrown with the vines of a wild rose bush, covered with countless blossoms. We sat down in this arbor of flowers and watched the sun rise over Herberhausen, submerging the small town in cascades of light. We did not move, we wanted to absorb the preciousness of these minutes completely. A new life was now spread out in front of us; it was as if the twilight had been conquered, the radiance surrounding us seemed like an omen.

The life, however, that was to begin now was not as rosy as that morning. It took another few months before we could move from Hechingen to Göttingen together with the whole Institute. But once there, our lives gradually stabilized more and more. The material problems could be dealt with; the care packages we received from friendly people throughout the world were a splendid help. And one thing was definitely different from beforehand: there was a future and hope again, and Heisenberg could work on his science again just as he desired, and this was exhilarating for him.

Karl Wirtz had taken over the task of building up the workshop and the experimental section of the Institute. In 1947, Professor Biermann from Hamburg could be obtained for astrophysics, and C.F. von Weizsäcker and Heisenberg were responsible for the theory. The main topic of those days was cosmic radiation — a harmless area, but full of fascinating possibilities. Reimer Lüst had returned from American internment, and the Institute was increasingly filling up with interesting and enthusiastic young people; Houtermans, Haxel, Koppe, Häfele, Schlüter, and many others; a highly talented and interested new generation was taking up its position. Most of them had already gone through a great deal, and were starved for objective, intensive scientific work, and they were full of new ideas and energy.

In the Mediterranean, a group supervised by Klaus Gottstein launched balloons from a chartered Italian gunboat to obtain pictures of cosmic rays. And later, under Reimar Lüst, these adventurous expeditions went as far afield as the Sahara. The results were evaluated back at home in Göttingen. Thus, a few years after the end of the war, a thriving Institute had blossomed out of the empty rooms of the AVA; indeed, it was scarcely able to contain all the people who had gathered there and spread out with their scientific activities. The festivities that were celebrated at the Institute during those years make for an eloquent description of a free, happy community, of pleasure in life, inexhaustible richness of ideas, and of a basically democratic attitude in which serious criticism was allowed to mingle with wit and humor. A few years later, Hans Peter Dürr returned from America, and together with him Heisenberg once again displayed a highly intensive scientific activity.

In the family, as well, things stabilized more and more. We had a nice house close to the woods, surrounded by a beautiful garden. A brisk life flourished there. Joint excursions into the delightful countryside around Göttingen on Sundays together with our friends frequently turned into large undertakings. Joint festivities were the high points. Once we were even able to find a magician, who bewitched 50 children in our house for a whole afternoon. The theater, under the ingenious direction of Heinz Hilpert, and the concerts conducted by Lehmann were of the highest quality. But above all, we enjoyed playing chamber music with friends, professional musicians and amateurs alike, at home; in time, the children could join in as well. And thus, what we had seen as a vision over the bench in Herberahausen came to pass: it turned into a happy time for us, and especially for Heisenberg, so that shortly before his death, he could say, "The time in Göttingen, it was the happiest time of my life."

Naturally, there were also disappointments, setbacks, and grievances during this period. Heisenberg was never really lucky in his dealings with politics. He had made up his mind to realize the political ideas he had carried around for so long. They had assumed a very concrete shape. He thought that a panel of 24 selected, responsible scientists should be formed that would be appointed to the government as a scientific council. It would act as an advisory, but also as a critical aid. He believed that scientists and politicians were by nature opposite types of people, who could fruitfully complement each other in political practice. The politician, he thought, was an active, vigorous, and decisive person, whereas the scientist could better be characterized by his high training and critical capacity for judgment; possibly the scientist could also be more creative than the politician, and not quite as bound by institutional forms. In his book, Heisenberg writes: "Obviously, it cannot be assumed that physicists and technicians could make better politically important decisions than politicians. But their scientific work has taught them to be objective and factual, and, what is more important, to keep the wider context in view. Hence they may introduce a constructive element of logical precision into the work of the politicians, of greater wisdom, and of objective incorruptibility that could be advantageous for this work." Heisenberg was convinced that the modern age that had started with the atom bomb could no longer survive without the far-seeing specialist. He writes about this in his book, as well: "The penetration of scientific thought, especially natural scientific thought, into the work of government was especially important to me. I thought that those who take on the responsibility for the functioning of the state needed a constant reminder that they were not dealing only with the balancing of conflicting interests. They needed to be reminded that there are frequently factually rooted necessities that are inherent in the structure of the modern state, and an irrational escape into emotionally colored thinking can only lead to catastrophe.

Heisenberg hoped that such a symbiosis between scientists and politicians would teach the German people a pragmatic way of thinking they were not exactly familiar with, and that this could prevent the repetition of extreme developments. He engaged all the energy not expended on his work in realizing this goal. His main partner in this effort was the Göttingen physiologist, Professor Rein.

Initially, developments looked promising, and the scientific advisory council, with Heisenberg as its chairman, was founded on March 9, 1949. But gradually it grew apparent that far greater resistance had to be overcome than he had thought. Simply assembling such a panel of scientists provoked annoyance and disapproval. And the officials in the ministries regarded the interference and the influence of the professors, who did not orient themselves toward practical considerations, as unacceptable, and boycotted the plan. In general, the forces of restoration gained more and more dominance in the body politic. There was no desire for experiments or risks, of which the outcome was unknown; there had been enough of that. Many thought it much safer and easier to fall back on the period before the Nazi regime; and so, in time, the notion prevailed that it was better to clearly separate politics and science.

Heisenberg could not overcome so much resistance. He already had a great amount of work with the reconstruction of the Institute and the intensive developments taking place there. Aside from that, his own new scientific ideas of a "unified field theory" were beginning to take tangible shape for the first time during this period, using up much of his energy. Actually, he had neither the time nor the strength successfully to overcome the forces that were now forming to oppose his intentions. Aside from that, tactics were foreign to him. Weizsäcker once expressed it this way: "The reasoning that the better argument, and not the tactics, always wins, as is the case in the sciences, made him the loser in political confrontations." Thus the "scientific advisory council" was finally dissolved, and the "research association" was founded in its place. In accordance with the model of the old exigency association of the 20's, it directed

26 Heisenberg around 1955

27 Niels Bohr, Werner Heisenberg, and Paul Dirac during a meeting of
Nobel laureates in Lindau (1962)

28 Carl Friedrich von Weizsäcker and Heisenberg at a meeting of the
Max Planck Society, around 1970

29 Werner and Elisabeth Heisenberg during a meeting of the Alexander
von Humboldt Foundation, around 1972

30 Heisenberg among his Humboldt scholarship recipients

31 The "international family of physicists" at the memorial celebration
for Niels Bohr in Copenhagen (1963)

Of the scientists mentioned in this book we find here: (from left to right)
In the first row the second is Professor Felix Bloch who had been
Heisenberg's assistant as a young scientist and later had to emigrate; next
to him Professor Fritz Hund. The eighth is Professor Victor Weisskopf,
next to him Professor Aage Bohr who took over the father's institute
after his death; next to Bohr Professor Paul Dirac with whom Heisenberg
travelled around the world in 1927. The thirteenth is Heisenberg, next to
him Professor Blackett who contacted Heisenberg during the imprisonment in England. Seated beside him is the mathematician Courant, also
an emigrant from Göttingen, who founded the famous Courant Institute
in New York.
In the second row the ninth is Professor Goudsmit, the eleventh Professor Wergeland, who is the only survivor of the three friends
Grönblohm, Euler, and Wergeland.
In the third row the third is Dr. Böggild whom Heisenberg managed to
have released from German imprisonment. The eleventh is Professor
C.F. von Weizsäcker, and next to him Professor John Wheeler.
In the middle of row four stands Professor Rozental with whom I had
correspondence.
Above him in row five Professor Casimir who has been the director of
the Philips AG for a long time.
Many of all the others whom I cannot list here counted among
Heisenberg's scientist friends.

its attention solely and exclusively toward matters of science. True, it was written in the statutes that the research association would aid the government with advice — this much Heisenberg had been able to achieve — but the practice was different, and reality saw the reinstatement of the old forms of absolute separation of the two great forces. Their tight fusion, Heisenberg's hope for a more humane and rational climate for German politics, was not to be. He not only regretted this development in politics, he also thought it to be dangerous for the sciences. He writes: "I explained my fears to X, that the exigency association he preferred could once again enhance the thought pattern that closes itself off from the hard, real world in an ivory tower, and indulges in fond dreams." That could not be allowed to happen! Approximately ten years later, circumstances demanded an organization in Bonn that at least partially realized Heisenberg's ideas. The creation of the Ministry for Research with its advisory councils brought about an institution representing a connection link between politics and science.

Heisenberg had fought for the realization of his ideas for so long and so hard that he was being accused of stubbornness and intractability. But finally, he had had to give in. He was very disappointed by this, for the political involvement he had shouldered with so much responsibility had been one of the main reasons for his decision to stay in Germany and to repeatedly turn down offers from abroad. It had been a hard blow.

But in spite of this severe setback, he did not withdraw from public life. He continued to remain active in many scientific-political undertakings. His strongest political interest was now directed at the common establishment [1954] and construction of the large international research laboratory in Geneva called CERN. He became Germany's chief representative during this difficult initial phase, when the goal was to embody the international character of this

large research facility in its constitution, while simultaneously defending German interests with tact and assurance. Frequently, this was no simple task. After the Institute
had been completed, he was asked if he would be willing
to take over the scientific supervision of CERN for five
years. He wavered for a long time. International work was
a strong temptation; another, just as strong, was the idea
that our children would grow up in the international framework of such a large Institution. Finally, he declined. There
was still too much to be done in Germany. To help in
reestablishing the international standard in physics after
decades of isolation was still one of the great challenges
to which he had subscribed. But if he went to Geneva for
five years, there was not much he could do to help. In addition, and perhaps this was decisive, his main interest was
directed at his own scientific work and ideas, and his
"general field theory," as it was later called, confronted him
with complicated and difficult problems, and he knew that
the work at CERN would not leave him any time for them.
As much of an honor as the offer was, and as much as
the challenge tempted him, he could not accept it.

His decision not to go to Geneva did not leave Heisenberg with the freedom to pursue his own work he had probably expected. There were too many tasks he did not think
he could refuse. In the meantime, his own Institute had
grown to a respectable size, and there was really no one
who could have replaced him. He was needed in the Max
Planck Society, in the Senate, and as an advisor with whom
Hahn could discuss his problems; he accepted the position
of President of the Götting Academy at its 300th year jubilee
celebration — in other words, when it had to represent the
continuity of scientific life in western Germany; he was
active in the Science Council in a leading position; time and
again, he had to represent Germany at international scientific conferences, and in 1953 he was offered the position
of President of the Alexander von Humboldt Foundation,

which was in need of rebuilding and reshaping. In none of these positions was he satisfied to be a mere representative. He always took on the tasks with his whole intensity. That he had to perform so many functions had a simple reason: there was a general lack of distinguished personalities. The utter devastation in all branches of our lives, brought about by terror, murder, displacement, and war, could be seen everywhere with terrible clarity. A new generation had to grow before it was possible to pass on the many duties. That started later, when Heisenberg had moved to Munich with his Institute.

chapter nine

BEARING THE RESPONSIBILITY

The dropping of the bomb on Hiroshima always struck Heisenberg as the beginning of a new era that would now be irrevocably overshadowed for all time by the terrible possibility of the explosion of an atomic bomb, an explosion that would cause unimaginable misery for millions of people. Naturally it was painful for him to think that what the physicists, himself included, had started in the 20's with such high spirits — namely, the penetration of the secret of the smallest particles of matter so as to understand their structure — had led to such a horrifying weapon of destruction, of unparalleled proportions. Had this made them guilty? Was science guilty? In his book, *Physics and Beyond*, he dedicates a whole chapter to the question of the responsibility of the researcher. In it he describes the walk he took with Carl Friedrich von Weizsäcker on the large field the morning after the bomb was dropped; they discussed the question of the guilt and responsibility the researcher bears in such a development. Let me quote some of the central ideas that were put forth: "This development [of modern science] is a process of life to which mankind, or at least European mankind, agreed centuries ago...We know from experience that this process can lead to good or evil. But we were convinced — and this is especially the belief in progress of the 19th century — that the good would prevail as knowledge grew, and that the possible evil consequences could be controlled. Prior to Hahn's discovery, neither he nor any one else could seriously consider the possibility of an atom bomb, as physics showed no way leading to it at the time. It cannot be regarded as guilt to participate in this life process of the development of science."

This was Heisenberg's conviction, and the idea that any new insight changes the world for the better as well as for the worse continually came to the fore in our frequent discussions of this question, and he saw any insight as a simultaneous challenge to the ethical conscience of mankind, especially to the politicians who mainly control the fate of nations. Further on, Heisenberg has Weizsäcker discuss the question of responsibility once more:

"It is probably necessary," he has Weizsäcker say, "to make a basic distinction between the discoverer and the inventor. As a rule, the discoverer cannot make any predictions about the possibilities of use prior to the discovery, and even afterwards the path to practical exploitation can be too long to permit any predictions...But in general, it seems to be different with the inventor...The inventor...has a certain practical goal in mind. He must be convinced that reaching that goal has its merits. Hence he must justly be burdened with the responsibility."

Once again, the statement is quite clear: insight, i.e. expansion of knowledge, cannot be associated with guilt, but invention, i.e. a development targeted at a goal, is within the responsibility of the researchers, the politicians, yes, even within the ethical quality of human society as such. The horror of the atomic bomb is obvious; but Heisenberg also believed that it offered a certain chance, the chance that it would awaken people, that it would sharpen our conscience and sensitize our feeling of responsibility, that, perhaps, the atomic bomb would prevent a new and terrible war with international dimensions. That was the hope I heard him speak about so often; that was the challenge to the human conscience. But Heisenberg was also increasingly tortured by the thought that scientific insights and new possibilities were developing too quickly, and that the ethical feeling of responsibility was not advancing at the necessary rate.

Heisenberg's political thoughts hardly fit into any political pattern. "You know," he would occasionally say to me, "politics always go this way or that, up and down. But if you yourself stay true to your course, then you are seldom 'right'; what I mean is that your own opinions will only rarely coincide with the official preaching of those having the 'say' at the moment." This was not said entirely without bitterness. Heisenberg was of the opinion that Germany should now once and for all forego any political attempt to achieve a position of power, be it economical or military, in Europe or even the world. In politics as well, Germany should stay in the background. As an example he used the Scandinavian countries and Switzerland, where through moderation, war had been avoided for so many centuries, and where life was so pleasant; this he had learned in Denmark. By the same token, however, sweeping declarations of peace should be avoided; he thought them to be worthless. "It is clear," he would say on occasion, "that the people's desire for peace is consistently misused by politicians for their own purposes. But we should do all we can to educate the people for peace. From the beginning, they have to be taught to forego something, even if they are convinced that they are in the right. All great conflicts," he said, "originate when each of the opponents believes himself to be in the right. And peaceful resolutions of conflicts can only become possible when we learn to think like the opponent, when we accept that the claims of the other side, seen from his standpoint, are correct. To see and to respect this, to accept this right of the other, is the precondition to striving for a balance of interests, and to even be able, for the sake of peace, to relinquish one's own rights. That is the beginning," he thought.

Having this attitude, it was a great shock for Heisenberg when he found out that the Federal Army of Germany was to be armed with nuclear weapons. This seemed to indicate that the development was once again going in the wrong direction. The initial horror people felt about the atomic bomb and its terrible effect had slowly given way to a kind of casualness, and there were certain tendencies to minimize the effects of the bomb, as though you could protect yourself from its deathly power in a bunker or cellar. Adenauer, one had the impression, nurtured this head-in-the-sand policy, for he thought that this NATO demand could not be turned down in the interest of German defense. But the responsible German nuclear physicists, combined in the "nuclear commission" chaired by Heisenberg, recognized the great risk inherent in such a policy. For, in case of a conflict between the super powers — it was the time of the "Cold War" between America and Russia — Germany would, without doubt, be the first target of hostile atomic bombs. They all thought the risk to be too high, and they felt that they had now suddenly all been placed in a position of political responsibility. The first to voice his concern was Weizsäcker. Heisenberg followed suit. Then the nuclear commission addressed the government in a letter, but it fell on deaf ears. The physicists did not relent, and soon the discussion of the question was raging on both sides. Adenauer and Franz Josef Strauss, then the Minister of Defense, were the proponents on the political side, and Heisenberg, Hahn, Weizsäcker, and Gerlach were the principal spokesmen on the scientific side.

In the midst of these controversies that went on for weeks, Heisenberg became very ill, and Weizsäcker became the spokesman of the discussions. When it grew obvious that Adenauer was not willing to change his mind in spite of all of the objections of the physicists, the decision was made to address the public directly. Weizsäcker and Heisenberg, who was sick in bed, worked out a manifesto directed against arming the Federal Army with nuclear explosives. The manifesto was signed by 18 leading nuclear physicists and appeared in German newspapers on April 13, 1957. The reverberations were worldwide. Examined closely, the manifesto was a consequent continuation of the

32 In a discussion with Hans Peter Dürr

33 In the seventies

34 Hans Peter Dürr, Heisenberg's colleague for many years, in Munich

35 Carl Friedrich von Weizsäcker during his time in Göttingen

36 In front of the Urfelder house

37 Heisenberg in his study in the Munich institute

38 At his lecture "What is an elementary particle?" during the annual meeting of the Physical Society in 1975

39 During the discussion following this lecture

opinion responsible physicists had held during the war. At the time, it had been born out of the resistance to an amoral, criminal government; now, it was the persuasion of accepted political impotence, endeavoring to solve political problems not by force, but through negotiations and understanding, and, if unavoidable, through necessary compromises and the renunciation of one's own rights, for the sake of peace.

But more than all the other things Heisenberg had committed himself to, it was his science that fulfilled him during these years. He was happy if he could work on his problems; he worked with his full concentration, and he worked to the point of exhaustion if the problem he was thinking about seemed intractable. Time and again, he was gripped by results that appeared to open up gates to new insights. One moonlit night we walked all over the Hainberg Mountain, and he was completely enthralled by the visions he had, trying to explain his newest discovery to me. He talked about the miracle of symmetry as the original archetype of creation, about harmony, about the beauty of simplicity, and its inner truth. It was a high point of our lives.

His emotion is revealed even more plainly in a letter he wrote to my sister, Edith Kuby, in January of 1958: "Dear Edith! I think it most charming of you to have thought of sending me this glorious bouquet of carnations. Thus, the happiness at success in work is augmented by the joy that others can emotionally participate in what has initially developed in a hidden scientific corner here. In fact, the last few weeks were full of excitement for me. And perhaps I can best illustrate what I have experienced through the analogy that I have attempted an as yet unknown ascent to the fundamental peak of atomic theory, with great efforts during the last five years. And now, with the peak directly ahead of me, the whole terrain of interrelationships in atomic theory is suddenly and clearly spread out before my eyes. That these interrelationships display, in all their mathematical abstraction, an incredible degree of simplicity, is a gift we can only accept humbly. Not even Plato could have believed them to be so beautiful. For these interrelationships cannot be invented; they have been there since the creation of the world. Naturally, there are still many details that need to be solved; but there will be many other people to help now, and that is good, as my own energies are not inexhaustible.

At the time, I must have written to my sister that Heisenberg was on the trail of such beautiful and exciting things, for his letter is, indeed, still full of happiness and optimism, but it is also still relaxed and peaceful; the storm broke out later through the indiscretion of a journalist, who, unknown to Heisenberg, participated in a university colloquium, where Heisenberg reported on his work at the request of Fritz Hund. The next day a sensational article was published, talking about the "world formula" offering the key to all still unsolved physical problems. This made the news in all small and large newspapers, both foreign and domestic. At first, Heisenberg thought he could simply ignore the whole uproar, but that was a mistake, bringing about many sad misunderstandings with his colleagues at home and abroad, who thought that he identified himself with these newspaper reports. In a letter to Pauli he complains about it: "During the last few days, I have had a lot of trouble with the newspapers. At the Institute, I had already lectured about our work a few times; there had been no consequences. Then Hund had asked me to talk about it at a more official university colloquium, as well. An incredible number of people showed up, and, without my knowledge, apparently some journalists. They published some hair-raising nonsense, along the lines of "the end of physics," etc. Hundreds of phone calls followed, and finally I dictated a couple of sentences to my secretary that she was authorized to quote as my opinion; the most impor-

tant of them was that our work was 'offering new suggestions for a unified field theory, the correctness of which
the research of the next few years will decide upon.' Following this, the nonsense subsided somewhat; but then, Landau must have stoked the fires of journalistic enthusiasm
in Moscow. In any case, reference was made to Landau's
speech in Moscow and things got even worse while I was
in Geneva. I hope you were not as annoyed as I! . . ."

The hope expressed in the letter to my sister, that "others
might now help," went unfulfilled. Heisenberg could not
comprehend why "all the others" were not willing to collaborate. But he was really deeply affected by the disapproval of Pauli, his old friend and life-long critic. He was
accustomed to presenting all his new ideas to him, but this
time he met with strong repudiation. And with this, the
passionate exchange of letters between Heisenberg and
Pauli, later called "the battle of Ascona," was initiated. At
the time, we were in Ascona to cure the viral encephalitis
that had so suddenly overcome Heisenberg during his
political altercation with Adenauer in April of 1957. There
were still intense periods of weakness and depressive phases,
but there was no thought of relaxation. For weeks the two
friends fought with each other in letters; if possible, each
letter was answered the same day. The letters were harsh
and without mercy. It was really like a battle, and each
volley was answered by an equally strong one from the
other side. This "battle" turned out well in the first round.
He finally succeeded in convincing Pauli of his ideas. Pauli
joined Heisenberg's side, and his letters displayed actually
euphoric statements. We were preparing to travel to Zürich.
Shortly thereafter, Pauli intended to go to Berkeley to give
lectures throughout the summer. Heisenberg was shocked
by the idea and advised him to postpone the trip for a while.
"As yet, you are not familiar enough with these new ideas,"
he told him that evening, "you will not be able to withstand the pressure of the Americans. You still have to

prepare all the necessary arguments." Thus he implored him to delay his trip. Pauli went anyway, and what Heisenberg had predicted took place. This was once again followed by embittered scientific quarrels. The conflict was resolved only while Pauli was in Varenna, shortly before his death. "You're free to go your own way," Pauli told him, "but I want nothing more to do with it." That is how they parted.

His illness marked a deep incision into Heisenberg's life, a clear warning. He had reached the limit of his strength, and he never completely regained his former ability. His radiant character was frequently overshadowed by fatigue and sadness, and the responsibilities he had formerly taken on so gladly were now becoming a burden to him. Nevertheless, his move to Munich gave him a fresh stimulus.

Naturally, there were many diverse reasons for the move to Munich. Last but not least, and he freely admitted this, was his long-standing desire to be able to live in Munich, after all — in Munich, the city he loved, where he felt alive and young.

Naturally, there were also more important reasons. The Institute in Göttingen had grown to such an extent that the old rooms of the AVA no longer offered nearly enough space. And Göttingen itself was a small city; the university was expanding to accommodate the flood of students. There was no question about it: Göttingen was too small for two so large and important, and even similar, institutions. They were severely crowding each other. Basically, the Max Planck Society belonged in Berlin, where it had always been located. But there was no longer any access to Berlin; the western occupying powers did not allow it, and Berlin was actually too exposed. On the other hand, Munich did not seem to be a bad solution. In the meantime its population had reached more than a million, and, among other things, it defended its title as the "hidden capitol" by extending offers to the Max Planck Society, and especially to Heisenberg.

In the end, though, a third argument proved decisive. Through lengthy negotiations with the Americans, mainly conducted by Heisenberg, it was resolved that Germany could once again carry on peaceful nuclear research. At first, the idea was simply to concentrate on a scientific test reactor, and a plan was devised to locate it in Garching, so that work on it could be executed in close collaboration with the Institute.

Heisenberg gave a lot of thought to the extent of participation such a problematic undertaking should receive, and in the chapter on the responsibility of the researcher in his book *Physics and Beyond*, he takes a stand. He writes: "In today's world, the life of man largely rests on this development of science. If we were to suddenly turn our backs on the continuous expansion of knowledge, the number of people inhabiting the earth would have to be reduced drastically in a short time. That could only be the work of a catastrophe comparable to, or even worse than, the atomic bomb." At the time, the problems of peaceful exploitation of nuclear energy were not clearly understandable. Heisenberg was convinced that the peaceful utilization of nuclear energy could be a blessing for mankind, and he was of the firm belief that the technicians would be able to solve all the problems arising from it. The recognition that we are approaching the limits of the possible in all areas is a historical process, and was not clearly visible in Heisenberg's time. Nevertheless, he had already written the following in his book *The Physicist's Conception of Nature** in 1955: "Hopes that the extension of man's material and spiritual powers would always spell progress are limited by this situation, if at first somewhat vaguely, and the dangers increase as the optimistic wave of faith in progress dashes against this limitation. Perhaps we might illustrate this kind of danger by means of an analogy. In what ap-

---

The Physicist's Conception of Nature, New York: Harcourt, Brace and Company, 1958. Translated into English by Arnold J. Pomerans.

pears to be its unlimited development of material powers, humanity finds itself in the position of a captain whose ship has been built so strongly of steel and iron that the magnetic needle of its compass no longer responds to anything but the iron structures of the ship; it no longer points north. The ship can no longer be steered to reach any goal, but will go round in circles, a victim of wind and currents. However, the danger persists only so long as the captain has not grasped that the compass is not responding to the magnetic forces of the earth. The moment he realizes that the danger is as good as half-removed; the captain who does not wish to sail in circles but wishes to reach a known or even unknown goal will find ways and means of determining the direction of the ship. He may use a modern compass which is not affected by the iron of the ship, or, as in the olden times, he may use the stars as his guides. Of course, he cannot order the stars to be visible at all times, and perhaps it is true that in our age only a few of them seem to be shining at all, but this one thing is clear: the very realization that faith in progress must have a limitation involves the wish to cease going in circles and to reach a goal instead. As we become clearer about this limitation, the limitation itself may be considered to be the first foothold from which we may re-orient ourselves." And a bit further on he writes: "This could imply that over longer periods of time a conscious acceptance of this limitation might well lead to some equilibrium, where man's knowledge and creative forces will once again find themselves ranged spontaneously about their common center." The test reactor was never built in Garching. Adenauer decreed — as he explained, because of French pressure — that the test reactor was to be the first step of a large nuclear research center in Karlsruhe. And so, the "reactor group," working under Karl Wirtz in the Institute, was split off and stationed in Karlsruhe. Heisenberg now had to decide whether he should go to Karlsruhe with his Institute, or whether he should agree with the split and move to Munich. The decision was not terribly difficult. To be sure, he had helped in the development of the basics of reactor construction during the war, but now responsibility was being passed on to experimental physicists and technicians, and that was fine with him. Thus the research center in Karlsruhe was the first large offshoot of Heisenberg's Institute.

A second split took place at the same time. Carl Friedrich von Weizsäcker accepted an offer from the philosophy department of Hamburg University, and thus did not go along to Munich. Heisenberg, however, moved to Munich with his family on January 25, 1958, and the official dedication of the Institute took place two years later. It had been built by his boyhood friend Sep Ruf.

Another period of fruitful work, dealing with many highly topical scientific and technical problems, unfolded in the attractive, bright building of the Institute. Heisenberg himself continued to work on his "general field theory" steadily and with great intensity, together with his excellent colleague and personal friend, Hans-Peter Dürr. In spite of this, he did not succeed in obtaining a breakthrough to general recognition and the cooperation of international physicists. The resistance he met was perhaps not always motivated by facts, but that is not our subject here.

The work of the Institute in Munich expanded into ever widening circles. From here, groups would travel to CERN, to experiment with the large machine. And new branches, new Institutes, grew out of the original one which had not been planned as large, and they took on a life of their own in Garching. The time of maturity and independence of the students had come.

It was a lucky coincidence that Heisenberg was offered a political task very appropriate to his talents at the time of his unsuccessful involvement in German national politics during the 50's. It was the Alexander von Humboldt Foundation, which he was to head for 22 years, until the time of his death. He made the Foundation into a political tool, corresponding to his ideas of the reconciling and internationally uniting force of science. The concept of the Foundation is to offer young, highly qualified scientists the opportunity to further their own studies in Germany, or to work on a topic that has to be researched specifically in Germany. The recipients of the awards are permitted to bring along their families and stay in Germany for one or even two years. Then, however — and this is a stipulation — they must return to their countries of origin. The selection of the recipients is based solely on their scientific qualifications; and it is the pride of the Foundation that world views, political convictions, religion, or race have no influence whatever on the selection decision. That is what makes the Foundation such a fertile and interesting meeting-place, so that common prejudices and biases can be dispersed. Under the leadership of Heisenberg and Dr. Heinrich Pfeiffer, the General Secretary, the Foundation blossomed during this period. Five thousand scientists from more than 80 countries were supported with research grants. Many of these former recipients are now active in leading positions and are attached to their host country, Germany, in friendship and gratitude.

Even when a recipient has returned to his country of origin, the Foundation maintains contact. Heisenberg always viewed it as a special duty to continue to offer a helping and protective hand. Thus he was able to step in and save the lives of the South African freedom fighter Alexander and of a group of Korean recipients when their political activities in their countries had brought them into a life-threatening conflict with their governments. Thus, in a certain sense, Heisenberg was living the continuation of his first political act during the revolutionary council period in Munich. Just as he had saved the life of the worker, then, whom he had guarded for a whole night, throughout his life he always felt responsible for "his" recipients, for whom he had taken on a small obligation as the president of the Foundation. It was not important whether the recipient was being persecuted from the "left" or the "right"; the only thing that mattered was that an industrious and suffering person was on the verge of perishing in the millstones of a political conflict; that was his sole motivation for doing everything in his power to try and save the person involved. So, in a certain sense, the Alexander von Humboldt Foundation became the fulfillment of the old dream he had believed in for so long, the dream of the "international family of scientists on the whole world," and the great resonance of the recipients gave him pleasure.

CONCLUSION

When we consider the course of Heisenberg's life, the inner consistency of his political thoughts and actions becomes overwhelmingly lucid. Everything interrelates with astounding clarity, and the motivations of his actions are apparent to the observer. He was never dominated by foreign or even friendly influences; he always acted on his own convictions, on his own conscience, and on his own perception of responsibility.

In the actual sense, Heisenberg was never "national," if this is understood to mean the irrational exaggeration of nation, in that one's own country is seen to have a higher value than all others, possessing overriding rights. He never thought that. But he loved the country of his childhood, and he felt both a part of it and responsible to it. And yet, essentially he saw himself as a European, especially and decisively shaped by the Anglo-Saxon world through his happy years in Copenhagen. In one of the first letters I received from him from Germany following his internment, i.e., from Alswede, written on January 1, 1946, he says: "All in all, I am doing quite well, as I have the feeling I am working for the future. Not only for our small circles, but for our larger cultural community. I would be pleased if in future this larger cultural community were called not only Germany, but Europe, but unfortunately, politics does not always do what one would like."

To some extent, Heisenberg's social involvement was privatized, in the sense that anyone to whom he felt committed, be it only because he had been asked for help, was somehow also "his neighbor," i.e., someone for whom he bore responsibility. This had the great advantage that he was accessible to everyone, regardless of which side the person was on. Thus his attitude was extremely humane, directed only toward the person. Heisenberg refused to brand one side as a criminal and the other as the champion of freedom and justice. He always saw the flaws of both sides, and with all violent arguments he feared the result could be a simple exchange of one misery for another, that there would only be a shift of the burden to a different class of people.

The difficulty with Heisenberg was that he did not think about politics as a politician but as a natural scientist. Just as he wanted to know how nature functions and how it is constructed, he wanted to know how politics is made and according to which laws it functions. He had an uncontrollable desire to learn and understand the laws governing the fate of man, of peoples, what controls world events. In this, his great teacher was Jacob Burckhardt, and his textbook was Burckhardt's *Weltgeschichtliche Betrachtungen*. This was the book he always reached for during the dangerous and oppressive period of the Nazi regime and the war, to understand what was happening around him. Burckhardt's carefully derived analysis of political forms — republic, aristocracy, monarchy, democracy, dictatorship — and especially, the chapter on "the crises," in which, using the course of the great, earth-shattering revolutions of antiquity as well as the events of the French Revolution, clearly demonstrate that the course of a revolution is always subject to certain general laws; and all this told Heisenberg that part ofthe events and the criminal development taking place before his very eyes, seemingly without any hope of their being interrupted, could be shifted into the objective realm. The agreement of Jacob Burckhardt's almost prophetic analysis with the events so oppressive to Heisenberg was a powerful solace to him. It gave him the chance to live with the development of events, with the decline of all values, of all morals in general.

His thirst for political knowledge was also the reason for his extreme interest in the people at the helm of politics. He never met Hitler or Himmler, and he never tried to. But he was attracted to and interested in encounters with men like Tito, Franco, Kennedy, de Gaulle, and Kissinger, and he enjoyed talking to people who he thought really did understand something about politics, like Adenauer, with whom he had many friendly discussions — even if they had occasional strong differences of opinion — or like Brandt, Helmut Schmidt, Marion Dönhoff, or, above all, Carl J. Burckhardt, with whom he enjoyed a warm friendship during the last ten years of his life.

And yet, there was an aspect in Heisenberg's political thinking that seemed utterly incomprehensible to many people and that gave rise to many misunderstandings. Heisenberg frequently regarded politics as a big game of chess, in which the feelings and passions of people are subordinated to the charted course of political events, just as chess figures to the rules of the game. And so, just as he was accustomed to playing complete games of chess in his head, without a board, he tried to think through political constellations as thought-experiments. His science had made him thoroughly familiar with thought-experiments; there he used them to discover the truth. And thus his political thinking frequently centered on complexes of ideas such as: power, force, criminal abuse, and the like. At times, he would then rather abruptly utter the result of one of his thought-experiments in a group, where the thought could do nothing other than cause extreme consternation. Thereby he occasionally inflicted harm on himself, for no one was capable of having even an inkling of the chain of ideas that had led to his hypothetical statement. Thus his listeners frequently had no other choice than to reach the wrong conclusions. And these highly abstracted speculations were then passed on, totally devoid of their context, by those he had unwittingly hurt or insulted, and they were grist for the mill of those who distrusted him politically or even wanted to harm him. And yet, if his thought processes were analyzed, it would probably be possible to recognize the pattern that always guided his political actions: to protect and care for the individual, helplessly at the mercy of the destructive forces of a world full of violence and crime.

Given this attitude, we can understand why he placed so little stock in revolutions, and even less in war. They brought immeasurable suffering to the individual, and in most cases they did not remove it. He thought in concepts of large spaces and long periods, and he frequently ran into contradictions with politicians and economists who did not understand him. Much later, his thoughts often showed themselves to have been quite close to the mark, and his proposals were then adopted in slightly altered form.

Heisenberg was never discouraged by his setbacks. And true to this resolution, until the end of his life he always made himself available when his advice or help was needed. Thus, in the final analysis, his political activity was extremely effective, although in a different way than he had envisioned, and his words were esteemed everywhere.

Obviously, we occasionally asked ourselves whether the decision to stay in Germany had been the correct one. But just as Heisenberg was not an unrealistic dreamer, he refused to muse about such senseless thoughts. The older he grew, the more his emotions flowed into the pleasure of living in this, in spite of it all, beautiful country, and thus he did take a positive stance toward his decision. For this reason he concludes his book with the following scene: "Von Holst fetched his violin, sat down between the two young men [our sons Wolfgang and Jochen] and joined them in playing the Serenade in D Major, a work of the young Beethoven, full of vital force and joy, and in which the trust in the central order of things overcomes all the faintheartedness and weariness everywhere." It was probably also his faintheartedness and weariness of the last few years of his life, during which he wrote his book, that were overcome by this trust, and by what he loved in Germany.

All that I have reported about Heisenberg is only a segment of his whole, rich personality. It says nothing about his winning friendliness, the patience with which he listened to others, nothing of his willingness to help whenever necessary, and to assist with thoughtful counsel, nothing of his almost shy modesty and the strong, yet controlled fervor with which he played the piano, nothing of his strength to keep things in limbo and to bear the tension of opposites. "You have to be able to stand living with the tension of opposites," he often said to me; that was the basis of his tolerant nature. And he continually measured his ethical standards by the book of Boethius, "The Consolation of Philosophy," which the latter, a high state official, had written in a dungeon some 1500 years ago when the tide of politics had turned once again. The language used in this book was his language, his way of thinking and of assessing the events of the time and his own life. Frequently I found him in his study holding this book, reading and reflecting, in a mood of calm concentration, yet open to spontaneous encounter.

I am not qualified to talk about him as a teacher and a scientist. He was committed to this side of his life with his greatest passion and intensity. With smiling certainty, he once said to me: "I was lucky enough to be allowed once to look over the good Lord's shoulder while He was at work." That was enough for him, more than enough! It gave him great joy, and the strength to meet the hostilities and misunderstandings he was subjected to in the world time and again with equanimity, and not to be led astray.

His spontaneous humanity was filled with great powers of conviction and warmth. It radiated forth and worked its magic on anyone meeting him without bias. There is so much evidence for this — last but not least, the overwhelming torchlight procession, during which all the members of the Institute placed candles in front of the room in which he had died. He had become a model for many people, and his simplicity and incorruptibility had won him a large following, and not only in Germany; no doubt, the fact that he had endured those difficult years in inner exile, and had not chosen his personal safety, had contributed to this.

Photographic Credits

Private collection of Mrs. Elisabeth Heisenberg: 1, 2, 3, 6, 7, 8, 9, 11, 12, 14, 15, 16, 17, 18, 19, 20, 23, 24, 31
Naturwissenschaften: 4
Fritz Hund: 5
G. Setti (Rome): 10
Edith Kuby: 13
Susanne Liebenthal (Frankfurt): 21
Transocean (Berlin): 22
nld Photo: 25
Erich Retzlaff: 26
dpa: 27
Max Planck Society: 28
Alexander von Humboldt Foundation: 29, 30
Gerhard Gronefeld (Munich): 32
Rita Strothjohann (Munich): 33
Max Planck Institute for Physics: 34
C. F. von Weizsäcker: 35
Christine Heisenberg-Mann: 36
W. Ernst Böhm (Ludwigshafen): 37
Margret Reiter (Munich): 38, 39

GLOSSARY

Biographical notes on some of the most important friends and colleagues of Heisenberg; annotations to several important institutions.

The numbers correspond to the numbers of the photographs.

KWS = Kaiser Wilhelm Society
KWI = Kaiser Wilhelm Institute
MPS = Max Planck Society
MPI = Max Planck Institute

Adenauer, Konrad 25

Alexander von Humboldt Foundation The Foundation was created as early as the 19th century for the advancement of the natural sciences, following the ideas of Alexander von Humboldt. It was newly founded in 1953. Its first President was Werner Heisenberg. The Foundation pursues regular contacts with over 6000 former visiting scientists in 83 countries. 29

Bagge, Erich Rudolf born 1912 in Neustadt (Coburg). Until 1948 at the KWI for Physics, then Professor in Hamburg, and in 1957 Prof. and Director of the Institute for fundamental and applied physics at the University of Kiel.

Beck, Ludwig Senior General, born 1880 in Bielnich, died 1944 in Berlin; Chief of the General Staff of the Army as of 1935. In 1938 he resigned and became the head of the resistance movement against Hitler.

Bethe, Hans born 1906 in Strasbourg, obtained his doctorate in 1928 under Sommerfeld, 1933 emigration; 1935 Professor at Cornell University; 1967 Nobel Prize for Physics.

Biermann, Ludwig born 1907 in Hamm/Westphalia; as of 1948 at the MPI for Physics in Göttingen. After the relocation of the Institute to Munich in 1958, he was the Director of the Institute for Astrophysics until 1977.

Blackett, Patrick b. 1897 in London, d. 1974; 1921-24 assistant of Ernest Rutherford; afterwards in Göttingen with James Franck. From 1925-33 with Rutherford again; 1933 Professor in London; 1927 in Manchester; 1953 he returned to London. 1948 Nobel Prize for Physics. 31

Bloch, Felix. b. 1905 in Zürich; obtained his doctorate in 1928 under Heisenberg; following some years as an assistant in Zürich (1928-29 with Pauli), Utrecht (1929-30 with Kramers), Leipzig (1930-31), and Copenhagen (1931-32), *Habilitation* in 1932 in Leipzig. In 1933 he went to the Henri Poincaré Institute in Paris and later to the University of Rome (to Fermi); 1934 he came to the United States and became an Assoc. Prof. (1934- 36) and Prof. (as of 1936) at Stanford University. 1954-55 General Director of CERN; Nobel Prize for Physics 1952. 31

Blount, Bertie b. 1907 in England; obtained his doctorate in chemistry (from Borsche in Frankfurt); after the war, the most important link between the scientists in Göttingen, the 'German Scientific Advisory Council', and the military government; 1950 Director of 'Scientific Intelligence' at the Ministry of Defense in London.

Böggild scientific colleague of Niels Bohr in Copenhagen. 31

Bohr, Niels b. 1885 in Copenhagen, d. there in 1962; studied physics in Copenhagen, where he obtained his doctorate in 1911. After various stays abroad he returned to Copenhagen and directed the Institute for Theoretical Physics in Copenhagen, a center for all natural scientific life, until his death. In 1922 he was distinguished with the Nobel Prize. 7, 9, 10, 14, 15, 27, 31

Bollmann, Erika in the general administration of the KWS since 1936; helped with the reconstruction of the KWS/MPS after the war; personal advisor of Prof. Butenandt during his term as president.

Bonhoeffer, Carl-Friedrich b. 1899 in Breslau, d. 1957 in Göttingen; obtained his doctorate in 1922 (from Nerust in Berlin); after a time as an assistant at the KWI in Berlin-Dahlem full professor in Frankfurt and Leipzig, where he was part of the closest circle of Heisenberg's colleagues. After the war Prof. in Berlin and Director of the Dahlem MPI for Physical Chemistry. In 1949 he became the Director of the MPI for Physical Chemistry in Göttingen.

Bonhoeffer, Dietrich theologian, b. 1906 in Breslau, murdered in the Flossenbürg concentration camp in 1945; 1935 Director of the Theological College of the Practicing Communion in Finkenwalde; was part of the resistance circle against the National Socialist government. He was arrested in 1943 and was later executed.

Born, Max b. 1822 in Berlin, d. 1970

in Bad Pyrmont; obtained his doctorate in 1906, *Habilitation* 1909 in Göttingen; Prof. in Berlin and Frankfurt. In 1921 he went to Göttingen where he engaged in great teaching and research activities. In 1933 he left Germany and received a professorship at the University of Edinburgh in 1936. After obtaining his emeritus in 1953, he returned to Germany and lived in Bad Pyrmont. In 1954 he received the Nobel Prize for Physics together with W. Bothe. 6

Bothe, Walther b. 1891 in Oranienburg, d. 1957 in Heidelberg; physicist, Prof. in Giessen and Heidelberg, Director of MPI for Medicine. Research in Heidelberg; 1954 Nobel Prize for Physics (with Max Born).

Burckhardt, Carl Jacob b. 1891 in Basel, d. 1978. Historian, writer, and diplomat; from 1937-39 High Deputy of the League of Nations in Danzig. From 1944-48 he was President of the International Red Cross, from 1945-49 envoy in Paris.

Burckhardt, Jacob b. 1818 in Basel, d. there 1897; studied theology, history, and art history (among others with Leopold von Ranke), 1855 professor in Zürich, 1885 in Basel.

Butenandt, Adolf b. 1903 in Wesermünde-Lehe, studied chemistry in Marburg and Göttinen; lecturer in Göttingen in 1931; 1933 received the Professorial Chair for Organic Chemistry at the Technical University in Danzig, in 1936 became the Director of the KWI for Biochemistry in Berlin. After the war he went to Tübingen. From 1960-72 he was president of the MPS. 1939 Nobel Prize for Chemistry.

Caratheodory, Constantin b. 1873 in Berlin, d. 1950 in Munich; Prof. of Mathematics in Munich.

Casimir, Gerhard b. 1909 in Den

Haag; studied physics and chemistry in Leiden, Göttingen, Copenhagen. In 1932 he went to the University of Berlin to Lise Meitner, then to Pauli in Zürich. From 1942 on with the Philips Company, where he became Executive Director. 31

CERN = *Conseil Européen pour la Recherche Nucléaire* (European Council for Nuclear Research), the European organization for elementary particle research was founded in Geneva in 1954.

Corinth, Lovis b. 1858 in East Prussia, d. 1925 in Berlin; painter and graphic artist; in 1919 Mrs. Corinth built the cottage bought by the Heisenberg family shortly before the war.

Debye (Debije), Peter Joseph b. 1884 in Maastricht, d. 1966 in Ithaca, N.Y.; *Habilitation* 1910 in Munich. Then professorships in Zürich, Utrecht, and Göttingen. From 1927-35 in Leipzig as a close colleague of Heisenberg. 1935 Director of the KWI for Physics in Berlin. 1940 he went to the United States. 1936 Nobel Prize for Chemistry. 10

Diebner, Kurt physicist, temporarily director of the Section for Nuclear physics in the Army Ordnance Office.

Diels, Ludwig b. 1874 in Hamburg, d. 1945 in Berlin, studied botany and became a professor in Berlin in 1921, as well as Director of the Botanical Garden in Berlin-Dahlem. He was a member of the *Mittwochgesellschaft*.

Dirac, Paul b. 1902 in Bristol; studied electrotechnology and mathematics in Bristol and Cambridge; from 1926-27 he was in Copenhagen with Niels Bohr. In 1927 he took a trip around the world with Heisenberg. Since 1932 Prof. at Cambridge University. 1933 Nobel Prize for Physics (with Erwin Schrödinger). 27, 31

Dönhoff, Countess Marion editor-in-chief since 1968, since 1973 publisher of *DIE ZEIT*.

Döpel, Gustav Robert b. 1895 in Neustadt/Orla; qualified for inauguration in physics in Würzburg in 1932 and became a professor in Leipzig. Worked on the initial form of the reactor with Heisenberg. From 1945-58 at research institutes in the Soviet Union; finally received a Professorial Chair at the University for Electrotechnology in Ilmenau.

Dolch, Heino b. 1912 near Leipzig; studied physics and theology and obtained his doctorate under Heisenberg. He belonged to the Jesuit Order. Qualified for inauguration in Münster. Subsequently Prof. in Paderborn and since 1963 in Bonn.

Dürr, Hans Peter b. 1929 in Stuttgart; studied experimental physics in Stuttgart. In 1953 he went to Berkeley, Calif. and obtained his doctorate from E. Teller in 1956. At Heisenberg's request he came to the MPI for Physics in 1957. Since then closest colleague of Heisenberg. Qualified for inauguration in Munich in 1970. Current Director of the MPI for Physics in Munich. 32, 34

Einstein, Albert b. 1879 in Ulm. d. 1955 in Princeton; studied physics in Zürich and Bern. After various other positions he became the Director of the KWI for Physics in 1914. In 1933 he emigrated and, after some time of roaming around, he received the executive position at the Institute for Advanced Studies in Princeton/U.S.A. 1921 Nobel Prize for Physics.

Euler, Hans studied theoretical physics in Leipzig and, after Bloch's emigration, became assistant to Heisenberg. Reported missing after a reconnaissance flight over the Soviet Union. 31

Glossary

Fermi, Enrico b. 1901 in Rome, d. 1954 in Chicago; studied in Pisa, in Göttingen (with Max Born), and in Leiden. Prof. in Florence, later in Rome. In 1939 he emigrated to the USA and received a post at Columbia University in New York, later a professorship in Chicago. He collaborated on the "Manhattan Project." 1938 Nobel Prize for Physics. 10

Flügge, Siegfried b. 1912 in Dresden, studied natural science in Dresden and Göttingen. Then a period of assistantship in Frankfurt and Leipzig, 1936-37, and at the KWI for Chemistry in Berlin-Dahlem. Professorships in Marburg and Freiburg.

Franck, James b. 1882 in Hamburg, d. 1964 in Göttingen; physicist, professor in Berlin, Göttingen, Baltimore, and Chicago; Nobel Prize for Physics in 1925.

Friedrichs, Kurt and Nellie Kurt Otto Friedrichs, b. 1901 in Kiel; studied mathematics in Göttingen, obtained his doctorate from Richard Courant in 1925, qualified for inauguration in Göttingen in 1929. Until 1937 professor at the Technical University in Braunschweig. Emigrated to the U.S.A. in 1937, where he received a professorship in 1943. From 1958-66 Director of the Courant Institute of Mathematical Science.
Nellie Friedrichs, nee Bruell, married to K.O. Friedrichs since 1937.

Gans, Richard Martin b. 1880 in Hannover, d. 1954 in City Bell, La Plata; studied electrical engineering, mathematics, and physics in Hannover and Strasbourg. In 1912 he went to La Plata, Argentina, as the Director of the Institute for Physics, returned to Germany in 1925. As of 1947 in Argentina again.

Geiger, Hans b. 1882 in Neustadt, d. 1945 in Berlin; pupil of Rutherford, Prof. in Kiel, Tübingen.

Gerlach, Walter b. 1889 in Bieberich am Rhein, d. 1979 in Munich; studied physics in Tübingen, where he qualified for inauguration in 1916. After years in the war and the industry he became Prof. for Experimental Physics in Frankfurt in 1920, Prof. in Munich in 1929 (as successor to Wilhelm Wien). He belonged to the group of ten nuclear scientists interned after the war. As of 1943 he was the deputy in the Research Council of the Reich for the German uranium project.

Gottstein, Klaus b. 1924 in Stettin; studied physics in Berlin, London, Göttingen. Since 1951 at the MPI for Physics; qualified for inauguration in Munich in 1960. 1971-74 scientific attaché at the German Embassy in Washington. 1974-80 at the MPI for Investigation of Vital Conditions of the Scientific-Technical world, Starnberg.

Goudsmit, Samuel b. 1902 in Den Haag, d. 1978 in Reno, Nevada; studied physics in Leiden; in 1928 he emigrated to the U.S.A. Various professorships in the U.S.A. From 1944-1945 Chief Scientific Officer of the "Alsos Project." 31

Grönblom, Berndt Olaf studied mathematics and physics in Finland; in 1935 he joined Pascual Jordan in Rostock and in 1936 he went to Leipzig. He died in the war between Russia and Finland. 31

Groves, Leslie R. b. 1896, d. 1979; American general. From 1942-47 he directed the "Manhattan Project."

Häfele, Wolfgang nuclear physicist, studied under Weizsäcker in Göttingen, then became the Director of the Institute for Applied Reactor Physics at the Nuclear Research Center in Karlsruhe.

Hahn, Otto b. 1879 in Frankfurt am Main, d. 1968 in Göttingen. He studied organic chemistry and turned to radiochemistry as a colleague of Rutherford in Montreal. Since 1912 Director of the radioactive division of the KWI for Chemistry in Berlin. From 1926 to 1942 Director of this Institute. As of 1946, President of the Max Planck Society. At the end of 1938, discovered nuclear fission and its chain reaction together with F. Strassmann. 1944 Nobel Prize for Chemistry 16, 25

Harteck, Paul b. 1902 in Vienna; studied chemistry in Vienna, Berlin and was Prof. for Physical Chemistry in Hamburg from 1934-51. Then he went to New York as a research professor.

Haxel, Otto b. 1909 in Neu-Ulm; studied physics in Munich and Tübingen, where he qualified for inauguration in 1936. Then he went to the Technical University in Berlin. From 1946-50 assistant at the MPI for Physics in Göttingen. 1950 professorship in Heidelberg.

Heisenberg, Annie nee Wecklein, b. 1871, d. 1945 in Bad Tölz; daughter of the Director of the Max Gymnasium in Munich. Married August Heisenberg in 1898. Two sons: Erwin, b. 1900, Werner Carl, b. 12-5-1901. 2

Heisenberg, August b. 1869 in Osnabrück, d. 1930 in Munich; as of 1888 he studied classical languages and byzantine history in Marburg, Munich, Leipzig, and Munich again. At first he was a teacher at various secondary schools in Munich, Lindau, and Würzburg, where he qualified for inauguration in Byzantine Studies while he carried out his duties in school. In 1911 he became the successor of his teacher, Krumbacher, in Munich and he expanded and enriched the field of research by including the cultural domain. 1, 2, 3

Heisenberg, August Wilhelm b. 1831, d. 1912 in Osnabrück; master locksmith.

Heisenberg, Erwin brother of

Werner Heisenberg. 2, 3

Heisenberg, Jochen b. 1939; studied experimental physics in Munich and Hamburg; obtained his doctorate in 1966. From 1967-69 he worked at Stanford with Hofstadter, thereafter at the MPI for Physics in Munich. From 1970-78 at MIT in Cambridge, Mass., since 1978 Prof. at the University of New Hampshire.

Heisenberg, Wolfgang b. 1938; studied law in Heidelberg and Munich. Now at the Thyssen Foundation in Cologne.

Hermann, Armin b. 1933 in Verona; studied natural sciences and qualified for inauguration in Munich in 1968; since 1968 Prof. for the History of Natural Sciences in Stuttgart.

Himmler, Gebhard headmaster of a secondary school, father of Heinrich Himmler.

Holst, Erich von b. 1908 in Riga, d. 1962 in Munich; zoologist, qualified for inauguration in Göttingen in 1934. 1946 Prof. in Heidelberg. 1949 Division Head of the MPI for behavioral physiology in Seewiesen.

Houtermans, Fritz Georg b. 1903 in Danzig; studied physics in Göttingen from 1921-27. Qualified for inauguration in Berlin and then went to the Soviet Union out of conviction in 1935; here he first worked in Charkow at the Ukrainian Institute for Physics, but later was caught in the web of the police and landed in prison. In 1940 he was successfully exchanged and returned to Berlin. In 1945 he went to Göttingen to the MPI for Physics; in 1952 he accepted an offer in Bern.

Humboldt Foundation see Alexander von Humboldt Foundation

Hund, Fritz b. 1896 in Karlsruhe; studied physics in Göttingen and Marburg. He obtained his doctorate and qualified for inauguration under Max Born in Göttingen.

Then two years in Copenhagen. In 1929 he went to Leipzig as a close colleague of Heisenberg. In 1956 he left the GDR and received a professorship in Göttingen. 31

Jacobi, Erwin b. 1884 in Zittau, d. 1965 in Leipzig; studied law and became an associate professor in 1912 at the University of Leipzig and a full Prof. in 1920. In 1933 he was relieved from his office and in 1945 he was reinstated. His fields were civil, administrative, labor, and canon law. He was among Heisenberg's closest friends.

Kaiser Wilhelm Society for the advancement of the sciences (see also Max Planck Society), founded in 1911 at the instigation of the then Kaiser Wilhelm II and upon the suggestion of Adolf von Harnack. Independent research institutes that were to support pure research only. The Society was supported mainly by private foundations and State grants. In 1946 the KWS was renamed Max Planck Institute. Its Presidents were:
Adolf von Harnack (1911-1930),
Max Planck (1930-1937),
Carl Bosch (1937-1940),
Albert Vögler (1940-1945),
Max Planck (1945-1946),
Otto Hahn (1946-1960),
Adolf Butenandt (1960-1972),
Reimar Lüst since 1972.

Kedar, Benjamin b. 1938 in Nitva/CSSR; studied medieval history in Jerusalem/Israel. Alexander von Humboldt grantee in 1976-77 at the German Institute for the Investigation of the Middle Ages, Munich. Now a lecturer at the Hebrew University of Jerusalem.

Kemble, Edwin b. 1889 in Delaware/Ohio; studied physics at various institutes in the U.S.A. Taught at Harvard University from 1919 to 1957. He is among the pioneers of quantum theory in the U.S.A.

Koppe, Heinz Walter b. 1918 in Leipzig; received his doctorate from Heisenberg in 1946. Until 1945 assistant at the MPI for Physics in Göttingen; then at the Universities of Heidelberg and Munich, since 1963 Prof. of Theoretical Physics at the University of Kiel. 20

Korsching, Horst b. 1912 in Danzig; studied chemistry, came from Otto Hahn's circle of colleagues to the KWI for Physics in 1937, later belonged to the MPI for Physics in Göttingen and Munich.

Kramers, Hendrik Antony b. 1894 in Rotterdam, d. 1952 in Oegstgeest/Holland; studied theoretical physics in Leiden and Copenhagen. He obtained his doctorate in 1919 and then worked at the Institute of Niels Bohr until 1926. Professorships in Utrecht, Leiden, Delft. 1947-51 President of the International Union for Pure and Applied Physics. 9

Kuby, Edith nee Schumacher, b. 1910 in Bonn; sister of Elisabeth Heisenberg, married the writer Erich Kuby in 1938.

Laue, Max von b. 1879 near Koblenz, d. 1960 in Berlin; studied mathematics, physics, and chemistry in Strasbourg, Göttingen, Munich, and Berlin, where he obtained his doctorate from Max Planck in 1903 and where he qualified for inauguration in 1906. From 1909-12 associate professor in Munich, then in Zürich and Frankfurt. In 1919 he went to Berlin and became a close colleague of Max Planck. In 1944 he went to Hechingen with Heisenberg's Institute. After the war he was interned in England; from 1946 on at Heisenberg's Institute at Göttingen. In 1951 he took charge of the Fritz Haber Institute in Berlin. 1914 Nobel Prize for Physics.

Lenard, Philipp b. 1862 in Pressburg, d. 1947 in Messelhausen; studied physics in Budapest, Vienna, Berlin, and Heidelberg, where he obtained his doctorate in 1886. In 1894 he qualified for inauguration under Heinrich Hertz in Bonn. After various professorships, he became a full professor for physics at the University of Heidelberg. In 1905 Nobel Prize for Physics. In the 20's he joined the NSDAP and became the main representative of the so-called "arian" or "German" physics.

Lindemann, Ferdinand b. 1852 in Hannover, d. 1939 in Munich; obtained his doctorate under Felix Klein in Erlangen in 1873; in 1877 he became associate professor and professor in Würzburg, in 1883 a professor of mathematics in Königsberg, in 1893 at the University of Munich.

Lüst, Reimar b. 1923 in Wuppertal-Barmen. After serving in the Navy and becoming an American prisoner of war, he studied physics in Göttingen. He qualified for inauguration at the University of Munich in 1960 and became a scientific associate and a section head at the MPI for extraterrestrial physics in the same year. Since 1972 President of the Max Planck Society.

Max Planck Society (see Kaiser Wilhelm Society)

Nachmansohn, David b. 1899 in Jekaterinoslaw/Russia; studied biochemistry and medicine (1926 M.D.) at the University of Berlin and then worked at the KWI for Biology (1926-30), at the Sorbonne (1933-39), and as a professor at the Yale School of Medicine (1939-42) and at Columbia College of Physiology and Surgery (after 1942). Since 1954 Prof. of biochemistry.

Ossietzky, Carl von b. 1889, d. 1938 in a concentration camp. He was an ardent pacifist and worked for the German Peace Society as of 1920, then as an editor of the "Berliner Volkszeitung" and for the periodical "Das Tagebuch," and from 1926-33 he published the magazine "Die Weltbühne." He was arrested after the burning of the German Parliament and died in a concentration camp. In 1935 he received the Nobel Peace Prize.

Pauli, Wolfgang b. 1900 in Vienna, d. 1958 in Zürich; studied theoretical physics under Sommerfeld in Munich from 1918-21, then he went to Max Born in Göttingen, to Hamburg, and to Copenhagen (1922-23 with Niels Bohr); in 1924 he qualified for inauguration and then went to Zürich, where he stayed until his death. Only during the war did he take a leave (1941-45), when he went to the U.S.A. to teach at the Institute for Advanced Studies in Princeton. 1945 Nobel Prize for Physics. 9, 23

Pegram, George b. 1876 in North Carolina, d. 1958 in Pennsylvania; since 1909 professor at various universities in the U.S.A. During the war he was the chairman of the Columbia Commission on War Research.

Pfeiffer, Henrich b. 1927 in Weinberg/Hessen; General Secretary of the Alexander von Humboldt Foundation since 1956.

Planck, Erwin son of Max Planck.

Planck, Max b. 1858 in Kiel, d. 1947 in Göttingen; studied physics in Munich and Berlin. Qualified for inauguration in Munich in 1880, 1885 private professor for physics in Kiel, 1889 professor in Berlin, 1913 rector of the University of Berlin, 1918 Nobel Prize, 1930-37 President of the Kaiser Wilhelm Society, 1943 escape from Berlin; 1945 arrival in Göttingen. Once more accepts the management of the KWS. At his suggestion, Otto Hahn is elected the new president in 1946. 19

Popitz, Johannes b. 1884 in Leipzig,

executed in 1945 in Berlin-Plötzensee. He was a lawyer and active in the Ministry of Finance of the Reich since 1919. He became Minister of the Reich in 1932 under General von Schleicher and was the Prussian Minister of Finance from 1933-44. In 1944, he was arrested in connection with the assassination attempt on Hitler.

Rausch von Traubenberg, Heinrich b. 1880 in Estonia, d. 1944 in Hirschberg; studied in Leipzig, Freiburg, Würzburg, and obtained his doctorate from W. Wien in 1905. After several professorships he set up a private laboratory in Berlin in 1937.

Reichwein, Adolf b. 1898 in Bad Ems, executed 1944 in Berlin-Plötzensee. In 1930 he received a professorship for History and Civics in Halle an der Saale. He joined the S.P.D. and was dismissed in 1933. He then worked as a village school teacher and at the Folklore Museum in Berlin. As of 1942, he was a member of the Kreisau Circle. He was arrested in 1944 and sentenced to death.

Rein, Friedrich Hermann b. 1898 in Mitwitz, d. 1953 in Göttingen; studied medicine and natural sciences in Munich and Würzburg. 1924, M.D. 1926 qualified for inauguration in Freiburg, since 1932 in Göttingen where he was the rector of the university from 1946-48.

Robinson, Charles F. b. 1915 in Vernon/Texas, studied natural sciences at the California Institute of Technology, Pasadena, and was part of the technical staff of the Office for Scientific Research and Development from 1941-45.

Rozental, Stefan long-time colleague of Niels Bohr.

Sauerbruch, Ferdinand b. 1875 in Barmen, d. 1951 in Berlin; important surgeon, especially in the field of chest operations. Member of the *Mittwochgesellschaft.*

Scherrer, Paul b. 1890 in St. Gallen, Switzerland; studied in Zürich, Göttingen, became an associate professor in Göttingen in 1918. From 1920-60 professor at the Federal Technical University of Zürich.

Schlüter, Arnulf b. 1922 in Berlin; obtained his doctorate in Bonn in 1947, scientific assistant of the KWI-MPI for Physics since 1948, as of 1959 a scientific member. Since 1960, during Heisenberg's chairmanship, member of the scientific management of the Institute for Plasma Physics. 1965-73 successor of Heisenberg as the chairman of the scientific management of the MPI for Plasma Physics. 1965-73 successor of Heisenberg as the chairman of the scientific management of the MPI for Plasma Physics, to which he still belongs today. 20

Schrödinger, Erwin b. 1887 in Vienna, d. 1961 in Vienna; studied physics at the University of Vienna from 1906-10. He qualified for inauguration in Vienna in 1914, went to the University of Jena (to Max Wien) in 1920, then became an associate professor at the Technical University of Stuttgart and a full professor at the University of Breslau. 1921 Professorial Chair for Theoretical Physics at the University of Zürich, in 1927 he was the successor of Max Planck at the University of Berlin. 1933 left Germany and initially went to Oxford (1933-38) and then to the University of Graz (1936-38) and finally to the Institute for Advanced Studies in Dublin (1939-56). Then he returned to the University of Vienna. 1933 Nobel Prize for Physics (together with Paul Dirac).

Schumacher, Hermann Albert b. 1868 in Bremen, d. 1954 in Göttingen; political economist. Following long trips abroad and several years' activity in the Prussian Ministry of Public Works, he became an associate professor in Kiel in 1899 and a full professor in Bonn in 1904. He founded the Technical University for Trade in Cologne. In 1917 he was called to Berlin. His main interest was the development of the world economy. Married to Edith Zitelmann, five children, among them Elisabeth (later Heisenberg).

Sommerfeld, Arnold b. 1868 in Königsberg, d. 1951 in Munich; studied mathematics in Königsberg, where he obtained his doctorate under Lindemann in 1891. In 1895 he qualified for inauguration in Göttingen; after various professorships he occupied the Chair for Theoretical Physics in Munich in 1906 and founded a very famous 'school' there that generated a whole generation of good physicists. 4

Spranger, Eduard b. 1882 in Berlin, d. 1963 in Tübingen; philosopher and pedagogue. After qualifying for inauguration in Berlin (1909), Prof. in Leipzig (as of 1911), in Berlin (1920- 46), then in Tübingen. He was a member of the *Mittwochgesellschaft*.

Stark, Johannes b. 1874 in Lower Bavaria, d. 1957 in Traunstein; studied the natural sciences at the University of Munich. Private lecturer at various universities. In 1922 he retired and worked in his own laboratory. 1919 Nobel Prize for Physics. During the 20's he joined the circle around Lenard and became an outspoken member of the NSDAP. In 1933 he was made the President of the Physical-Technical Institute of the Reich, in 1939 the President of the German Research Association.

Strassmann, Fritz b. 1902 in Boppard, d. 1980 in Mainz; chemist, Prof. in Mainz, discovered uranium fission in 1938 together with O. Hahn.

Supec, Ivan b. 1915 in Zagreb; studied theoretical physics in Leipzig and obtained his doctorate in 1940. During the war he fought with the resistance under Tito. In 1945 he became Professor for Theoretical Physics at the University of Zagreb, founded a large institute and also had a significant political influence under Tito's rule, in the sense of helping to create a genuine democratization.

Teller, Edward b. 1908 in Budapest; studied physics at the Technical University of Karlsruhe and in Leipzig, where he obtained his doctorate from Heisenberg in 1930. Then Göttingen 1931-33, Copenhagen 1934, London 1935-41. Professor at the George Washington University in Washington; from 1941-51 he worked on the American atom and hydrogen bomb project, then in Chicago, and finally in Berkeley.

Telschow, Ernst b. 1889 in Berlin; studied chemistry, physics, technology in Munich and Berlin. Obtained his doctorate under Otto Hahn in 1911. In 1931 he entered into the general administration of the KWS and in 1937 became the managing administrator of the Society. In 1948 member of the Executive Council of the MPS, pensioned in 1954. 1954-59 manager of the Society for Physical Studies in Düsseldorf. 1959-68 construction of the MPI for Plasma Physics in ·Garching near Munich. Honorary Senator of the MPS.

Tetzler brother-in-law of Paul Dirac, Rumanian merchant.

Tomonaga, Sin-Tiro b. 1906 in Tokyo, d. there 1979; studied physics in Kyoto until 1929; worked with Heisenberg in Leipzig from 1937-39. Then he returned to Japan. In 1965 he received the Nobel Prize for Physics together with Richard Feynman and Julian Schwinger.

Glossary

Vögler, Albert b. 1877 in Barbeck, d. 1945 in Dortmund; was metallurgical engineer, industrialist, 1917-36 chairman of the Association of German Iron Workers, 1926 co-founder of the United Steelworks Inc., politician (1919-20 Weimar National Assembly, 1920-24 representative of the German "Volkspartei"). From 1941-45 president of the KWS.

Volz, Helmut b. 1911 in Göppingen, d. 1978 in Erlangen; studied physics in Tübingen, Munich, and Leipzig. 1944 professor at the University of Erlangen; worked in the field of nuclear physics.

Waerden, Barthel Lendert van der b. 1903 in Amsterdam; studied mathematics in Amsterdam, Göttingen, and Hamburg. Qualified for inauguration in Göttingen in 1927, in 1931 he accepted an offer in Leipzig, where he remained until 1945, and where he belonged to the circle of Heisenberg's trusted colleagues. After the war, occupied primarily with the history of mathematics, in Zürich.

Wecklein, Nicolaus b. 1843 in Gainsheim/Lower Franconia; d. 1926 in Munich, studied classical philology in Würzburg; qualified for inauguration in Munich, but then went to the Max Gymnasium as a teacher, and later to Bamberg, Passau, and in 1886 he became the rector of the Max Gymnasium in Munich.

Weisskopf, Victor b. 1908 in Vienna; studied theoretical physics in Vienna and Göttingen. Obtained his doctorate from Max Born in 1931. Afterwards a year in Leipzig, then Berlin. Finally assistant to Wolfgang Pauli in Zürich. 1936 a year in Copenhagen. Then he went to America. During the war he collaborated on the "Manhattan Project." Finally professor at MIT in Cambridge, Mass. From 1960-65 he

was the head of CERN in Geneva. 31

Weizsäcker, Carl Friedrich von b. 1912 in Kiel; studied physics in Berlin, Göttingen, Leipzig, where he obtained his doctorate under Heisenberg in 1933. Active at the KWI for Physics in Berlin as of 1936. From 1942-44 he taught in Strasbourg as an associate professor, then he returned to the Heisenberg Institute in Hechingen. As of 1946, section head at the MPI for Physics in Göttingen, in 1957 went to Hamburg as a Professor of Philosophy. 1969-80 Director of the MPI for Investigation of Vital Conditions of the Scientific-Technical World in Starnberg. Weizsäcker was among Heisenberg's closest friends. 28, 31, 35

Wergeland, Harald b. 1912; studied physics in Trondheim, Leipzig, Copenhagen, and Oslo and became Professor for Theoretical Physics at the Technical University of Trondheim in 1946. 31

Weyl, Hermann b. 1885 in Elmshorn; d. 1955 in Zürich; studied mathematics in Munich and Göttingen and obtained his doctorate under David Hilbert in 1908. Professorships in Göttingen and Zürich, successor to Hilbert in Göttingen. Left Germany in 1933 and went to Princeton.

Wheeler, John A. b. 1911 in Jacksonville, Florida; studied physics at the Johns Hopkins University, Baltimore, where he obtained his doctorate. From 1933-35 he was a Fellow of the National Research Council in New York and Copenhagen, from 1935-38 associate professor at the University of North Carolina, and since 1938 professor at Princeton University. He has been teaching at the University of Texas in Austin for the last three years. 31

Wieland, Heinrich b. 1877 in Pforzheim; d. 1957 in Starnberg;

studied chemistry in Berlin, Karlsruhe, and Munich. He qualified for inauguration at the University of Munich in 1904 and, with only one short interruption, he stayed there until the end of his life. Nobel Prize for Chemistry 1927.

Wien, Max b. 1866 in Königsberg, d. 1938 in Jena; physicist, professor in Danzig and Jena.

Wien, Wilhelm b. 1864 in East Prussia; d. 1928 in Munich; studied mathematics and physics in Göttingen and Berlin and obtained his doctorate under Helmholtz in Berlin in 1886. After various stops in Aachen, Giessen, and Würzburg, he received the Professorial Chair for Experimental Physics in Munich in 1920; he stayed there until his death.

Wirtz, Karl b. 1910 in Cologne, studied physics and physical chemistry in Bonn, Freiburg, and Breslau and in 1935 became an assistant of Bonhoeffer in Leipzig. In 1937 he joined the KWI for Physics in Berlin. He became Section Head in 1944; 1954 Director of the Institute for Neutron Physics and Reactor Technology in Karlsruhe; simultaneously also a professor at the Technical University. 24